# DOMAIN STRUCTURE IN FERROELECTRICS AND RELATED MATERIALS

# DOMAIN STRUCTURE IN FERROELECTRICS AND RELATED MATERIALS

*A.S. Sidorkin*

Viva Books

New Delhi | Mumbai | Chennai | Kolkata | Bangalore | Hyderabad | Kochi

**First Indian Edition 2008**

**VIVA BOOKS PRIVATE LIMITED**

- 4737/23 Ansari Road, Daryaganj, New Delhi 110 002
E-mail: vivadelhi@vivagroupindia.net, Tel. 42242200
- Plot No. 76, Service Industries, Shirvane, Sector 1, Nerul, Navi Mumbai 400 706
E-mail: vivamumbai@vivagroupindia.net, Tel. 27721273, 27721274
- Jamals Fazal Chambers, 26 Greams Road, Chennai 600 006
E-mail: vivachennai@vivagroupindia.net, Tel. 28290304, 28294241
- B-103, Jindal Towers, 21/1A/3 Darga Road, Kolkata 700 017
E-mail: vivakolkata@vivagroupindia.net, Tel. 22836381, 22816713
- 7, Sovereign Park Apartments, 56-58, K. R. Road, Basavanagudi, Bangalore 560 004
E-mail: vivabangalore@vivagroupindia.net, Tel. 26607409, 26607410
- 101-102 Mughal Marc Apartments, 3-4-637 to 641, Narayanguda, Hyderabad 500 029
E-mail: vivahyderabad@vivagroupindia.net, Tel. 27564481, 27564482
- Beevi Towers, First Floor, SRM Road, Kaloor, Kochi 682 018, Kerala
E-mail: vivakochi@vivagroupindia.net, Tel: 0484-2403055, 2403056

***www.vivagroupindia.com***

*Published by arrangement with*
Cambridge International Science Publishing
7, Meadow Walk, Great Abington
Cambridge CB1 6AZ, UK



ISBN-10: 81-309-0831-X
ISBN-13: 978-81-309-0831-1

Published by Vinod Vasishtha for Viva Books Private Limited, 4737/23, Ansari Road, New Delhi-110 002. Printed and bound by Anand Sons, Delhi-110 092.

Contents

# Introduction

An important place among the solid-state materials is occupied by dielectrics and the so-called 'active dielectrics' in particular. The latter have received their name because of their ability to manifest qualitatively new properties under the external influence. Pyroelectrics, piezoelectrics and ferroelectrics are traditionally considered as active dielectrics. Polarization switching at temperature variation is characteristic of pyroelectrics, onset of polarization under the action of mechanical pressure is peculiar to piezoelectrics. Both of the above-mentioned classes of the active dielectrics are linear dielectrics, i.e. they are the substances for which the effect taking place is proportional to ???the value of an action, and the value of the proportionality constant is permanent????.

A special place among the active dielectrics is occupied by ferroelectrics. These are the substances possessing spontaneous polarization in a definite temperature range, i.e. spontaneously occurring polarization, which can be reversed, in particular, by applying an external electric field to a crystal. The special significance of these materials is connected with the non-linearity of their properties, which enables their characteristics to be controlled with the help of external actions.

The fact of implementation of their polar states in the form of the so-called domain structure is one of the distinctive features of ferroelectric materials. An individual domain represents a macroscopic area in a crystal, in which, for instance, in ferroelectrics, all elementary cells are polarized in the same way. The directions of spontaneous polarization in the neighboring domains form certain angles with each other. A system of domains with different orientation of the polarization vector represents the domain structure.

Considerable attention devoted to such a seemingly individual material property as the domain structure is relevant to the fact that practically all main distinctive properties of ferroelectrics are interdependent. This means that their non-linear properties and complete switching processes as well as all other features are determined to a large degree by the state and mobility of the domain structure. Therefore, in order to study the nature of these properties and their possible applications in practice, it is crucial to find out the regularities that control the processes of origination of the domain structure and the ways of its change with time.

It is well known that the change of macroscopic polarization in ferroelectrics takes place by means of displacement of boundaries between domains. These boundaries are called domain walls. Therefore, studies of properties of domain structures cannot be separated from the investigation of processes of domain boundaries motion.

Ferroelastics are closely related to ferroelectrics as far as their properties are concerned. They are substances in which spontaneous deformation of elementary cells takes place at certain temperatures. The spontaneous deformation in ferroelastics as well as polarization in ferroelectrics occur at structural phase transformations. This also determines the likeness of methods of the theoretical description of these materials. These methods involve symmetry-related principles, studies of the properties of the corresponding thermodynamic functions, etc. That is why it is quite natural to consider simultaneously the properties of the mentioned ferroactive materials, the patterns of domain structure and its dynamics, wherever it is possible.

The ferroelectric materials possess a lot of useful applied properties. The presence of sustained polarization that lasts without the action of a field, for example, makes it possible to use them for recording and retrieving information. At the same time, the density of information storage in ferroelectrics is much higher as compared to magnetic media due to the significantly thinner transient layer (domain wall) between domains, which makes their utilization preferable from the point of view of at least this factor. Recently, a discussion was started about the possibility of utilization of the periodicity of the arrangement of domain walls for the generation of laser radiation with the required wavelength, etc.

Thus, the studies of the domain structure of ferroelectrics represent both fundamental and applied interest. In reality, considerable attention is devoted to this problem, which is reflected by the number of articles in magazines on general physics and by numerous scientific conferences, both international and Russian, etc. At the same time in contrast to ferromagnetics, for example, there is practically no monograph literature which would be devoted solely to that problem. The parts of the books devoted to the general properties of ferroelectrics and dealing with this problem are usually too brief and deal only with the experimental description of the domain structure [1–28]. Despite the analogy to the ferromagnetics a simple transfer of the results obtained for the ferromagnetics to the ferroelectrics is not possible. The characteristics of the domains and domain boundaries in ferroelectrics are controlled by the interactions that differ from the ones in ferromagnetics. That fact brings certain specifics. Namely, the width of the domain boundaries in ferroelectrics is several orders of magnitude smaller and, consequently, their interaction with the crystalline lattice and its defects is very strong. Their bending displacements are controlled not by surface tension but rather by long-range fields. The screening effect and its influence on the domain structure and on the domain walls motion do not have any analogues in ferromagnetics, etc.

This book represents an attempt to bridge the gap. It is devoted to the description of the main characteristic parameters of the domain structure and domain boundaries in ferroelectrics and related materials.

As probably in any publication, the problems considered in the book reflect, of course, certain preferences of the author. For example, the first chapter deals with the mechanisms of formation of the domain structure. The formation of the domain structure is studied most thoroughly in the framework of the mechanism of loss of initial phase stability in the finite size material. Particular attention is devoted to the equilibrium domain structure and the so-called fine domain structure. A hypothesis is analyzed that the origination of the fine domain structure is connected with the transition to a new phase under the conditions of inhomogeneous cooling of only a thin layer of a ferroelectric material. It is shown that this hypothesis can explain the fact of the onset of a periodical domain structure in a ferroelastic with a free surface.

In the second chapter, the structure of domain and interphase boundaries in defect-free ferroelectrics and related materials is considered within the framework of the phenomenological description of materials with different types of phase transitions. The influence of the concentration of charge carriers, material surface, etc. on the boundaries under consideration is examined. The problems of stability of different types of domain boundaries are discussed.

The third chapter presents the results of the microscopic description of the structure of domain boundaries in ferroelectrics. Ferroelectric crystals of barium titanate and of the potassium dihydrophosphate group are taken as an example. The results of the microscopic and phenomenological descriptions are compared. The limits of validity of the phenomenological way of description of the problem under consideration are assessed.

The fourth chapter deals with the description of the interaction of domain boundaries in ferroelectrics and ferroelastics with different types of crystalline lattice defects. The processes of interaction of domain boundaries in ferroelectrics and ferroelastics with various types of crystalline lattice defects are studied. This includes charged defects, dilatation centers, non-ferroelectric inclusions, dislocations with different orientation of Burgers' vector with respect to the direction of spontaneous shear and the domain wall plane.

The fifth chapter deals with the problems of stability of the shape of inclined domain boundaries, and also with the structure of domain boundaries in real ferroactive materials. The concept of the effective width of a domain wall is introduced. It is shown that the deformation of a domain wall shape in materials with defects can be the reason for domain wall widening in real materials.

In the sixth chapter, the influence of lattice potential relief on the mobility of domain walls is studied. The thermofluctuational mechanism of the motion of domain walls is considered, parameters and the probability of appearance of a critical nucleus on a domain wall is calculated. On the basis of the results of the given consideration the explanation to the effect of domain structure 'freezing' in ferroelectrics of the potassium dihydrophosphate group is given. The influence of the proton tunneling effect on hydrogen bonds on the structure and mobility of domain boundaries in ferroelectrics containing hydrogen is studied.

In the seventh chapter, the proper and forced dynamics of domain boundaries in ferroelectric and ferroelastic crystals are considered. Bending and translational dynamics of domain boundaries in ferroelectric crystals are studied; the contribution of domain boundaries to the dielectric properties of ferroelectrics and elastic properties of ferroelastics is investigated. The experimental data on the dielectric properties of ferroelectrics with different types and concentration of defects are analyzed. In the final part of this chapter the non-linear dielectric properties of ferroelectrics, associated with the motion of domain boundaries and processes of ageing and degradation of ferroelectric materials, are briefly considered.

Finally, I would like to thank very much Messrs. S.Kamshilin for helping with the proofreading and correction of the translation and K.Penskoy for his help in typesetting the book.

Chapter 1

# Formation of a domain structure as a result of the loss of stability of the crystalline lattice in ferroelectric and ferroelastic crystals of finite dimensions

When discussing the reasons for the formation of the domain structure, we usually emphasize the symmetric [7] and energy [4, 5] aspects. According to the Curie principle, the symmetry of crystal after influence is the result of multiplication of the symmetry of the crystal before influence by the symmetry of the influence itself. Since temperature is a scalar, then from the viewpoint of the theory of symmetry, at least macroscopically, as a result of the phase transition caused by, for example, a change of temperature, the symmetry of the crystal should not change. However, since the symmetry of the crystal decreases within the limits of each domain, to restore the symmetry in the material as a whole, structural changes in the given domain are balanced by the opposite changes in another domain.

The restriction of domain parameters in symmetric consideration is evidently the equality of only total volumes of domains of unlike sign. However, in reality, if we disregard the case of crystals with the so-called internal field [21], the equality of not only the average but also individual dimensions of domains is observed, i.e. a strictly periodic domain structure.

The strict periodicity of the domain structure is naturally linked with the minimization of the general energy of the system in such a domain structure. Since the size of the domains is finite, it is evident that the general balance of the minimized energy should contain terms with the opposite dependence on the width of the domain $d$. In the case of ferroelectrics in particular these terms are the energy of the depolarizing field of bound charges of spontaneous

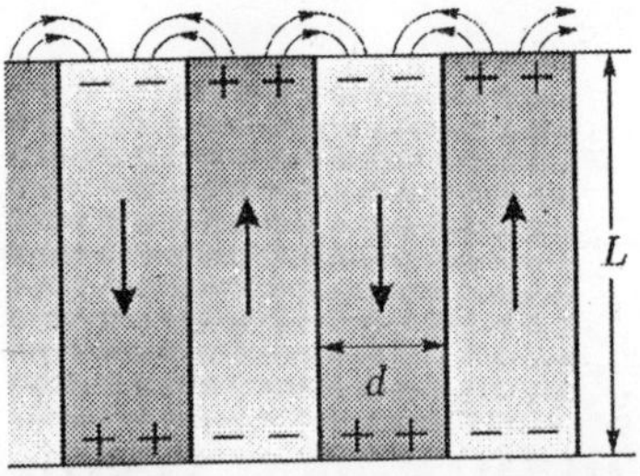


**Fig. 1.1.** Decrease of the energy of the depolarizing field of a ferroelectric specimen of finite dimensions after division into domains. $L$ is the size of the crystal along the polar axis, $d$ is the average domain width.

polarization on the surface of the crystal and correlation energy or the energy of domain boundaries. Their minimisation in particular for the 180° (laminated) domain structure, results in the so-called Kittle domain structure (Fig.1.1) described in the case of ferromagnetics for the first time by Landau and Lifshits [22], and in ferroelectrics by Mitsui and Furouchi [29]. Later on, the investigation of the equilibrium domain structure for the case of ferroelectric domains, mechanical twins and magnetic domains has been carried out in papers [30–33] and [34], respectively.

## 1.1 Equilibrium domain structure in ferroelectrics

Let us determine the width of a laminated equilibrium domain structure. For this purpose, let us first of all find the energy of the depolarizing field at an arbitrary ratio between the dimensions of the unlike sign domains $d_+$ and $d_-$ respectively, ignoring, as it is done in [35], the variation of spontaneous polarization along the polar axis $z$ in the vicinity of the ferroelectric surface $z = 0$. The thickness of the surface non-ferroelectric layer $\Delta$ will be assumed to be zero to simplify considerations.

The surface density of a charge in this case

$$\sigma(x) = \begin{cases} P_0, & 0 < x < d_+, \\ -P_0, & d_+ < x < d_+ + d_- = 2\bar{d}. \end{cases} \tag{1.1}$$

can be conveniently represented by a Fourier series

$$\sigma(x) = \frac{a_0}{2} + \sum_{n=1}^{\infty} \{a_n \cos\frac{\pi n x}{\bar{d}} + b_n \sin\frac{\pi n x}{\bar{d}}\}, \tag{1.2}$$

where

$$a_0 = \frac{P_0[d_+ - d_-]}{\bar{d}}, \quad a_{n\neq 0} = \frac{2P_0}{\pi n}\sin\frac{\pi n d_+}{\bar{d}},$$
$$b_n = \frac{2P_0}{\pi n}\left[1 - \cos\frac{\pi n d_+}{\bar{d}}\right], \tag{1.3}$$

$P_0$ is spontaneous polarization.

The electrical potential $\varphi$ satisfies the Laplace equation

$$\varepsilon_c \frac{\partial^2\varphi}{\partial z^2} + \varepsilon_a \frac{\partial^2\varphi}{\partial x^2} = 0 \tag{1.4}$$

($\varepsilon_c, \varepsilon_a$ are dielectric permittivities of the monodomain crystal along and across the polar axis) with boundary conditions

$$\left.\frac{\partial\varphi}{\partial z}\right|_{+0} - \varepsilon_c \left.\frac{\partial\varphi}{\partial z}\right|_{-0} = -4\pi\sigma(x), \quad \varphi_{+0} = \varphi_{-0}. \tag{1.5}$$

From (1.4), taking into account (1.5) and (1.2), we obtain

$$\begin{aligned}\varphi(z<0) &= \sum_n e^{\lambda_n z\sqrt{\frac{\varepsilon_a}{\varepsilon_b}}}\{A_n \cos\lambda_n x + B_n \sin\lambda_n x\} - \frac{\pi a_0 z}{\varepsilon_c},\\ \varphi(z>0) &= \sum_n e^{\lambda_n z}\{C_n \cos\lambda_n x + D_n \sin\lambda_n x\}\end{aligned} \tag{1.6}$$

where

$$\begin{aligned}&A_n = C_n, \quad B_n = D_n, \quad A_n = \frac{4\pi a_0}{\lambda_n(1+\sqrt{\varepsilon_c\varepsilon_a})},\\ &B_n = \frac{4\pi b_n}{\lambda_n(1+\sqrt{\varepsilon_c\varepsilon_a})}, \quad \lambda_n = \pi n/\bar{d}.\end{aligned} \tag{1.7}$$

Taking into account (1.3), the surface density of the energy of the depolarizing field

$$\Phi = \frac{1}{\bar{d}}\int_0^{\bar{d}} \varphi(x,0)\cdot\sigma(x)dx + \frac{a_0}{2}\varphi(z=-L), \tag{1.8}$$

where $L$ is the size of the crystal along the polar axis, is as follows:

$$\Phi = \frac{8P_0^2\bar{d}}{\pi^2(1+\sqrt{\varepsilon_c\varepsilon_a})}\sum_{n=1}^{\infty}\frac{1}{n^3}\left\{\sin^2\frac{\pi n d_+}{\bar{d}} - \left(1-\cos\frac{\pi n d_+}{\bar{d}}\right)^2\right\} + \frac{\pi P_0^2 L}{\varepsilon_c}\frac{[d_+ - d_-]^2}{2\bar{d}}. \tag{1.9}$$

As shown by further investigations, at any equilibrium size $\bar{d}$ (1.14), the formation of a unipolar structure, i.e. the structure with $d_+ \neq d_-$, increases $\Phi$. The minimum of $\Phi$ corresponds to the overall unpolarized structure. In this case

$$\Phi = \frac{16P_0^2 d}{\pi^2(1+\sqrt{\varepsilon_c\varepsilon_a})}\sum_{n=1}^{\infty}\frac{1}{(2n-1)^3}. \tag{1.10}$$

As it can be seen from (1.10), the energy of the depolarizing field decreases with the refining of the domain structure. Adding to (1.10) the total energy of the domain walls

$$\Phi_\gamma = \gamma\frac{L}{d} \tag{1.11}$$

($\gamma$ is the surface density of the energy of domain boundaries), with the inverse dependence on the domain size $d$, and minimizing the sum of (1.10) and (1.11), we obtain the following expression for the equilibrium size of a domain

$$d = \left( \frac{\pi^2 \gamma L \sqrt{\varepsilon_c \varepsilon_a}}{16.8 P_0^2} \right)^{1/2} \tag{1.12}$$

or taking into account the specific expression for γ:

$$\gamma = \frac{4}{3} \sqrt{\frac{\pi \varkappa}{\varepsilon_c}} P_0^2 \tag{1.13}$$

we have

$$d = \frac{\varepsilon_x^{1/4} \varkappa^{1/4} L^{1/2}}{\sqrt{10.2\pi}}. \tag{1.14}$$

The above minimization is carried out at a temperature corresponding to the observation conditions that reflects the equilibrium nature of the domain structure. At the same time, in experiments one often comes across a non-equilibrium domain structure the parameters of which are not determined by the observation conditions but by the conditions of formation of the domain structure. In our opinion the formation of a domain structure starts with the phase transition as a result of the loss of stability of the crystalline lattice in the phase transition to the low temperature phase in relation to the fluctuation of the order parameter with the value of the wave vector differing from zero.

In fact, the investigation of the phase transition in a ferroelectric crystal of finite dimensions shows that here in contrast to an infinite crystal takes place the transition to the state with the nonuniform distribution of polarization. Apparently, this state is a prototype of the subsequently formed domain structure in which the initial wave-shaped distribution of polarization is replaced by the step-like distribution with clearly defined domain boundaries together with the increase of polarization as a result of non-linear interactions. The mobility of these boundaries and, even to a greater extent, their number are restricted because of various reasons and, that's why in the observed domain structure only due to kinetic reasons, for example, one can expect the presence of the same period as in the initial distribution of polarization. In other words, as a first approximation, the size of the domain is taken here as the period of the modulated distribution of the order parameter, which occurs during the phase transition. The identification of these periods with each other enables to provide an accurate quantitative estimate for the period of the domain structure in ferroelectrics, [36, 37] and ferroelastics [38–41], and to explain the formation of the fine-domain structure [42], the very fact of appearance of a regular domain structure in ferroelastics with free surface [43], as well as to describe the variation of the period of the domain structure in ferroelectric materials with free charge carriers and charges on the surface layer [44, 45]. Evidently, the domain structure obt using suchin this approach is non-equilibrium because its formation conditions differ from the observation conditions.

## 1.2 Formation of a modulated structure in a ferroelectric crystal under the conditions of homogeneous cooling

Let us consider the main results of the proposed approach. We start with the case of a ferroelectric crystal in the form of a thin plate cut in the direction normal to the polar axis. It is assumed that the thickness of the ferroelectric material is $L$, it is surrounded by a surface non-ferroelectric layer of thickness $\Delta$ and dielectric permittivity $\varepsilon$ and is either placed or (in another case) not placed inside a shortened capacitor.

To determine the distribution of polarization formed during the phase transition in the crystal and the accompanying electric fields, we start with the simultaneous equations that include material equations for the ratio of the polarization components $P_x$ and $P_z$ along the non-polar axis $x$ and polar direction $z$ with the electrostatic potential

$$\alpha_x P_x = -\frac{\partial\varphi}{\partial x}, \qquad -\alpha_z P_z - \varkappa\frac{d^2 P_z}{dx^2} = -\frac{\partial\varphi}{\partial z} \tag{2.1}$$

and Laplace's equation

$$\varepsilon_x \frac{\partial^2\varphi}{\partial x^2} - \varepsilon_z \frac{\partial^2\varphi}{\partial z^2} = 0, \tag{2.2}$$

where $\varepsilon_x = 1 + 4\pi/\alpha_x$, $\varepsilon_x \approx 1 - 4\pi/(\alpha_z - \varkappa k^2)$, $-\alpha_z = \alpha_0(T - T_c)$, and $k$ is the wave vector in the wave dependence of $P_x$ and $\varphi$ on the coordinate $x$.

The distribution of the potential in different areas (Fig.1.2) in the absence of electrodes will be found in the form

$$\begin{aligned} \varphi_I &= Ae^{-kz}, \\ \varphi_{II} &= Be^{-kz} + Ce^{kz}, \\ \varphi_{III} &= D\sin\left(\sqrt{\varepsilon_x/\varepsilon_z}\,kz\right). \end{aligned} \tag{2.3}$$

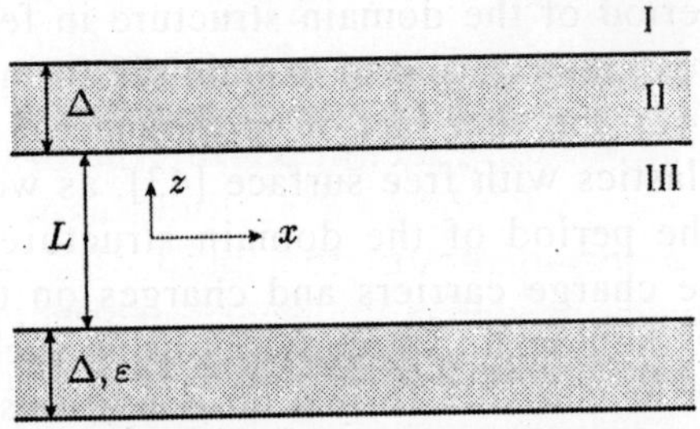


**Fig. 1.2.** Ferroelectric material with surface non-ferroelectric layers.

Solution (2.3) should satisfy the conditions of joining of the potential at the interfaces of the media I, II, III, and also the condition of continuity of the normal induction components

$$\varphi_{\mathrm{I}} = \varphi_{\mathrm{II}}\Big|_{z=\frac{L}{2}+\Delta}, \quad \varphi_{\mathrm{II}} = \varphi_{\mathrm{III}}\Big|_{z=\frac{L}{2}},$$

$$\varepsilon\frac{\partial\varphi_{\mathrm{II}}}{\partial z} = \frac{\partial\varphi_{\mathrm{I}}}{\partial z}\bigg|_{z=\frac{L}{2}+\Delta}, \quad \varepsilon\frac{\partial\varphi_{\mathrm{II}}}{\partial z} = -\varepsilon_z\frac{\partial\varphi_{\mathrm{III}}}{\partial z}\bigg|_{z=\frac{L}{2}}. \tag{2.4}$$

The simultaneous equations (2.4) allow to find the ratios between the unknown coefficients in the expressions for the potentials (2.3). The condition of the solvability of these equations, i.e. the equality to zero of the determinant, compiled from the coefficients of the quantities $A$, $B$, $C$, $D$, produces an equation determining the dependence of coefficient $\alpha_z$ on the wave vector $k$ modified taking into account the effect of correlation and electrostatic interaction of bound charges on the surface of the crystal.

Substitution of the distribution (2.3) into (2.4) and notation of the given determinant

$$\begin{vmatrix} e^{-k\left(\frac{L}{2}+\Delta\right)} & -e^{-k\left(\frac{L}{2}+\Delta\right)} & -e^{k\left(\frac{L}{2}+\Delta\right)} & 0 \\ -e^{-k\left(\frac{L}{2}+\Delta\right)} & \varepsilon e^{-k\left(\frac{L}{2}+\Delta\right)} & -\varepsilon e^{k\left(\frac{L}{2}+\Delta\right)} & 0 \\ 0 & e^{-k\frac{L}{2}} & e^{k\frac{L}{2}} & -\sin(tL/2) \\ 0 & e^{-k\frac{L}{2}} & -e^{k\frac{L}{2}} & -\dfrac{\sqrt{\varepsilon_x/\varepsilon_x}}{\varepsilon}\cos(tL/2) \end{vmatrix} = 0. \tag{2.5}$$

$$t = k\sqrt{\varepsilon_x/\varepsilon_x}$$

yields an equation for the link of $\alpha_z$ with $k$ of the following type:

$$\operatorname{tg} k\sqrt{\frac{\varepsilon_x}{\varepsilon_z}}\frac{L}{2} = \frac{\sqrt{\varepsilon_x\varepsilon_z}}{\varepsilon}\frac{[(\varepsilon+1)+(\varepsilon-1)\exp(-2k\Delta)]}{[(\varepsilon+1)-(\varepsilon-1)\exp(-2k\Delta)]}. \tag{2.6}$$

In the absence of a non-ferroelectric layer, i.e. at $\Delta \to \infty$ and $\varepsilon = 1$, it transforms to the equation [37, 46, 47]

$$\mathrm{tg}\, k\sqrt{\frac{\varepsilon_x}{\varepsilon_z}}\frac{L}{2} = \sqrt{\varepsilon_x \varepsilon_z}, \tag{2.7}$$

from which, taking into account the ratio of $\varepsilon_z(k)$ with $\alpha_z$, the approximate dependence $\alpha_z = \alpha_z(k)$ has the form of

$$\alpha_z(k) = -\alpha_z + \varkappa k^2 + \frac{4\pi^3}{\varepsilon_x^2 k^2 L^2}. \tag{2.8}$$

The loss of stability takes place with respect to such a value of the wave vector which corresponds to a minimum of $\alpha_z(k)$ dependence (see Fig.1.3), i.e. regarding the value of

$$k_m = \frac{\sqrt{2}\pi^{3/4}}{\varepsilon_x^{1/4} \varkappa^{1/4} L^{1/2}}. \tag{2.9}$$

Like the period of the equilibrium structure (1.14), the period of modulated distribution $d = \pi/k_m$ from (2.9) is proportional to $L^{1/2}$ and has the same dependence on other parameters. It is not surprising because quantity $d$ here is determined by the balance of the same interactions as in (1.14). However, even the difference in the type of periodic solution (step-like in (1.14) in comparison with sinusoidal in (2.9)), i.e. the presence in (1.14) of not one but of an entire set of harmonics, results in a quantitative difference

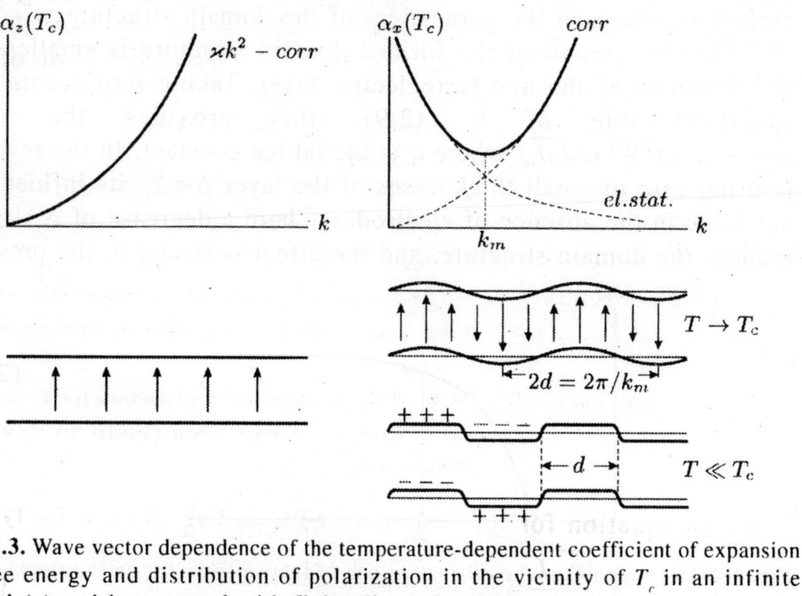


**Fig. 1.3.** Wave vector dependence of the temperature-dependent coefficient of expansion of free energy and distribution of polarization in the vicinity of $T_c$ in an infinite crystal (a) and in a crystal with finite dimensions (b).

of their periods by as much as four times. As it is estimated, only part of this difference can be attributed to ignoring the distribution of polarization along the polar axis when determining the period of the equilibrium domain structure, and, consequently, the solution of this section is metastable.

The substitution of the resultant value of the wave vector $k_m$ (2.9) in the dependence $\alpha_z(k)$ (2.8) makes it possible to find the temperature at which the transition to the state with the inhomogeneous distribution of polarization will take place. Its shift in relation to $T_c$ of the infinite crystal in the direction of low temperatures as a result of the overturning effect of the depolarizing field in relation to the onsetting polarization is equal to

$$\Delta T = \frac{4\pi^{3/2}\varkappa^{1/2}}{\alpha_0 \varepsilon_x^{1/2} L}. \tag{2.10}$$

If the specimen, subjected to a phase transition, is placed in a capacitor, the value of potential $\varphi_I = 0$. Consequently, the equation for determining the dependence $\alpha_z(k)$ is transformed here into the condition

$$\operatorname{tg} k\sqrt{\frac{\varepsilon_x}{\varepsilon_z}}\frac{L}{2} = \frac{\sqrt{\varepsilon_x \varepsilon_z}}{\varepsilon}\operatorname{th} k\Delta. \tag{2.11}$$

The analysis of the obtained equations (2.6) and (2.11) shows, that the presence of a surface ferroelectric layer with not too high dielectric permittivity both in the presence or in the absence of electrodes has almost no effect on the parameters of the domain structure at $k\Delta \gg 1$, i.e. when the period of the formed domain structure is smaller than the thickness of the non-ferroelectric layer. Taking into account the specific value of $k_m$ (2.9), this provides the value $\Delta = \Delta_1 \approx (\varkappa L^2)^{1/4} \approx \sqrt{aL}$, where $a$ is the lattice constant. In the reversed limiting case of small thicknesses of the layer $\Delta \ll \Delta_1$, its influence is not large in the absence of electrodes – here a decrease of $\Delta$ slightly reduces the domain structure, and the effect is strong in the presence

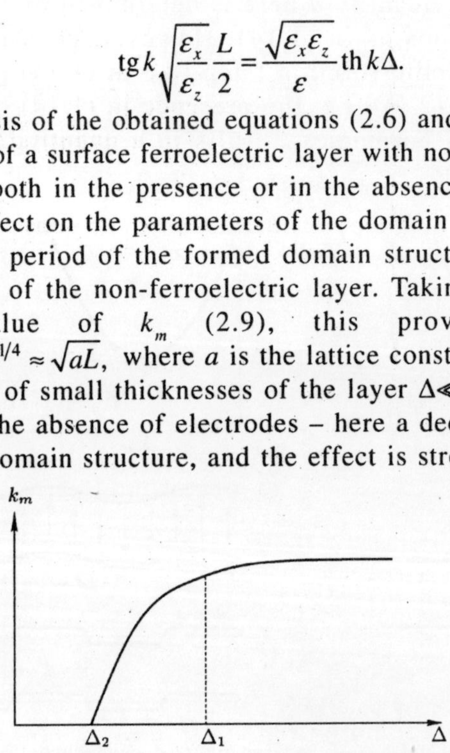


**Fig. 1.4.** Dependence of the critical values of the wave vector on the thickness of the surface non-ferroelectric layer.

of electrodes – here a decrease of $\Delta$ is accompanied by a large increase of the size of the domain structure up to the formation of a monodomain structure at $\Delta = \Delta_2 = (\varkappa\varepsilon / \pi^3 \varepsilon_x)^{1/2}$ (Fig.1.4).

The increase of the dielectric permittivity of the surface layer on its own in both cases increases the width of the domain as the result of additional dielectric screening.

## 1.3. Formation of the domain structure in a ferroelectric plate of an arbitrary cut

Evidently, another factor affecting the parameters of the domain structure formed during a phase transition may obviously be the orientation of the ferroelectric plate in relation to the polar axis, i.e. the type of cut used in practice. In fact, for any so-called 'skew' cut, i.e the cut for which the vector of spontaneous polarization in the ferroelectric plate is not perpendicular to its surface, the density of the bound charge of spontaneous polarization on the plate with other conditions being equal is smaller in comparison with the straight cut. At the same time, a skew cut plate is characterised by an increase of the length of domain boundaries. Consequently, the balance of energy factors, determining the width of the domain, changes here, evidently in such a manner that in the skew cut plate one should expect an increase of the distance between adjacent domain walls with the increase of the angle of deviation of the spontaneous polarization vector from the direction normal to the surface of the plate.

For the quantitative description of the above effect let us consider a case of an arbitrary skew cut. Let us place the laboratory system of coordinates $xyz$ in such a manner that the $z$ axis is normal to the plane of the plate of the uniaxial ferroelectric, and the $y$ axis is normal to the axis of spontaneous polarization. The angle between the $z$ axis and the axis of spontaneous polarization or ferroelectric axis is denoted by $\psi$. The crystallographic system of coordinates $x'y'z'$ has the axis $z'$ parallel to the axis of spontaneous polarization $P_0$; the directions of the axis $y$ and $y'$ coincide.

To derive equations, determining the distribution of polarization and the electric potential in the considered system let us write in advance the thermodynamic potential of the plate in the crystallographic system of coordinates

$$\Phi = \int_V \left\{ \frac{\alpha_x}{2} P_x^2 - \frac{\alpha_z}{2} P_z^2 + \frac{\varkappa}{2} \left( \frac{\partial P_z}{\partial x'} \right)^2 + \frac{E^2}{8\pi} \right\} dV. \qquad (3.1)$$

Here, as previously, $\alpha_z = \alpha_0(T_c - T)$, $\mathbf{E} = (E_x, E_z)$ is the depolarizing field. This problem can be conveniently solved in the laboratory system of coordinates. The transition from the crystallographic to laboratory system of coordinates is determined by the following correlations:

$$x' = x\cos\psi - z\sin\psi,$$
$$z' = x\sin\psi + z\cos\psi, \tag{3.2}$$
$$y' = y.$$

From this it follows that

$$\frac{\partial}{\partial x'} = \frac{\partial}{\partial x}\cos\psi - \frac{\partial}{\partial z}\sin\psi,$$
$$\frac{\partial}{\partial z'} = \frac{\partial}{\partial x}\sin\psi + \frac{\partial}{\partial z}\cos\psi, \tag{3.3}$$
$$\frac{\partial}{\partial y'} = \frac{\partial}{\partial y}.$$

The minimization of thermodynamic potential (3.1) in respect of the components of the polarization vector leads to simultaneous equations which, taking into account (3.3), may be expressed as follows

$$\alpha_x P_z = E'_x = -\frac{\partial\varphi}{\partial x'} = -\frac{\partial\varphi}{\partial x}\cos\psi + \frac{\partial\varphi}{\partial z}\sin\psi, \tag{3.4}$$

$$-\alpha_z P_z - \varkappa\frac{\partial^2 P_z}{\partial x'^2} = E'_x = -\frac{\partial\varphi}{\partial z'} = -\left(\frac{\partial\varphi}{\partial x}\sin\psi + \frac{\partial\varphi}{\partial z}\cos\psi\right). \tag{3.5}$$

Supplementing equations (3.4), (3.5) with the electrostatic equation

$$\Delta\varphi = 4\pi\nabla P \tag{3.6}$$

yields a complete set of equations for determination of the components of the vectors $\mathbf{P} = (P_x, P_z)$ and $\mathbf{E} = (E_x, E_z)$.

Taking into account the linearity of equations of the set (3.4-3.6) with respect to the components of the polarization vector **P** and the potential $\varphi$, the solution of this set in our case may be presented in the form

$$\varphi = \varphi_0 \exp(ikx)\exp(itz), \tag{3.7}$$

where $k$ is the real wave vector and $t$ is a parameter determining the variation of $\varphi$ in the thickness of the plate. It may be complex in general.

Substituting (3.7) into (3.4) and (3.5) we find

$$P_x = \frac{i}{\alpha_x}(-k\cos\psi + \lambda\sin\psi)\varphi_0 \exp(ikx)\exp(itz), \tag{3.8}$$

$$P_z = \frac{i(k\sin\psi + t\cos\psi)}{\alpha_z - \varkappa k^2\cos^2\psi}\varphi_0\exp(ikx)\exp(itz). \tag{3.9}$$

Substituting subsequently (3.8) and (3.9) into (3.6) taking into account (3.3) in the approximation $\psi < 1$, we obtain the following expression for the relationship between $t$ and $k$:

$$t_{1,2} = \frac{4\pi\sin\psi\cos\psi(\alpha_x + \alpha_z) \pm \alpha_x\alpha_z\sqrt{\left(\frac{4\pi}{\alpha_z} - 1\right)\left(\frac{4\pi}{\alpha_x} + 1\right)}}{\alpha_x\tilde{\alpha}_z\left[1 + 4\pi\left(\frac{\sin^2\psi}{\alpha_x} - \frac{\cos^2\psi}{\tilde{\alpha}_x}\right)\right]}k, \tag{3.10}$$

where

$$\tilde{\alpha}_z = \alpha_z - \varkappa k^2\cos^2\psi. \tag{3.11}$$

In a special case of $\psi = 0$ ratio (3.10) yields formula (2.5)

$$t = \pm\sqrt{\frac{\varepsilon_x}{\varepsilon_z}}k \tag{3.12}$$

of the previous section where

$$\varepsilon_x = 1 + \frac{4\pi}{\alpha_x}, \quad \varepsilon_z = \frac{4\pi}{\tilde{\alpha}_z} - 1. \tag{3.13}$$

The solution of simultaneous equations of the equilibrium for polarization with the potential (3.7), noted down with consideration of two roots in (3.10), in the form

$$\varphi = A\exp(ikx)\exp(it_1z) - B\exp(ikx)\exp(it_2z), \quad |z| < \frac{L}{2}, \tag{3.14}$$

$$\varphi = C\exp(ikx)\exp(-kz), \quad |z| > \frac{L}{2}, \tag{3.15}$$

must satisfy the boundary conditions on the surface of the plate at $z = \pm L/2$:

$$\varphi(x,z)\big|_{x=\frac{L}{2}-0} = \varphi(x,z)\big|_{x=\frac{L}{2}+0}, \tag{3.16}$$

$$\left.\frac{\partial\varphi}{\partial z}\right|_{z=\frac{L}{2}-0} - \left.\frac{\partial\varphi}{\partial z}\right|_{z=\frac{L}{2}+0} = 4\pi P\Big|_{z=\frac{L}{2}-0}, \tag{3.17}$$

where $P = P_z\cos\psi - P_x\sin\psi$ is the projection of polarization normal to the surface of the plate, that onsets in the specimen as the result of phase transition.

From (3.16) and (3.17) in the approximation $4\pi/\tilde{\alpha}_z \gg 1$ we find an equation determining the dependence of $\alpha_z$ on $k$:

$$\sqrt{\varepsilon_x \frac{4\pi}{\tilde{\alpha}_z}}\ \mathrm{tg}(t_1 - t_2)\frac{L}{2} = 2. \tag{3.18}$$

Since in the vicinity of the phase transition point the value of $\tilde{\alpha}_z$ is small, equation (3.18) may be rewritten in the form

$$(t_1 - t_2)\frac{L}{2} = \pi n, \tag{3.19}$$

where $n$ is any integer. On the basis of (3.10) we find from (3.19)

$$\sqrt{\varepsilon_x \tilde{\alpha}_z} = \frac{-2\pi^{3/2} n \cos^2\psi}{kL}. \tag{3.20}$$

Analysis of (3.20) shows that here the value $n = 0$ corresponds to the onset of the homogeneous state of the ferroelectric phase, whereas $n = -1$ corresponds to the stable state of the formed heterogeneous ferroelectric phase. From (3.20) the required dependence of $\alpha_z$ on $k$ in the approximation of $kL/2 \gg 1$ is

$$\alpha_z(k) = -\alpha_z + \varkappa k^2 \cos^2\psi + \frac{4\pi^3 \cos^4\psi}{\varepsilon_x k^2 L^2}. \tag{3.21}$$

The minimum of the dependence (3.21) corresponds to the value

$$k_m = \frac{\sqrt{2}\pi^{3/4}}{(\varepsilon_x \varkappa)^{1/4} L^{1/2}} \cos^{1/2}\psi, \tag{3.22}$$

which at $\psi = 0$ changes to $k_m$ (2.9) for the straight cut. Thus, the period of the domain structure formed here [48] is

$$d = \frac{\pi}{k_m} = \frac{(\varepsilon_x \varkappa \pi)^{1/4} L^{1/2}}{\sqrt{2}\cos^{1/2}\psi}. \tag{3.23}$$

Substituting $k_m$ (3.22) in (3.21) and equating $\alpha_z(k) = 0$, we obtain that the transition to the state with the heterogeneous (modulated) distribution of polarization in the skew cut plate is shifted in relation to the $T_c$ of the infinite crystal in the direction of lower temperatures by the value

$$\Delta T = \frac{4\pi^{3/2} \varkappa^{1/2}}{\alpha_0 \varepsilon_x^{1/2} L} \cos^3\psi. \tag{3.24}$$

As expected, in accordance with (3.23) the period of the onsetting domain structure may be controlled by selecting the appropriate orientation of the cut of the ferroelectric plate.

## 1.4. Formation of the domain structure under the conditions of polarization screening by charges on surface states and by free charge carriers

The period of polarization distribution formed in phase transitions, which becomes subsequently the period of a metastable domain structure changes greatly in the presence of polarization screening as well that strongly influences not only the parameters but also the type of domain structure [49]. This screening may be implemented by both charges on the surface state and by free charge carriers.

The mechanism of the influence of spontaneous polarization screening on the equilibrium width of the domain represents the decrease of the energy of the depolarizing field. In this case, for the balance of the energies of the depolarizing field and the domain boundaries that takes place in equilibrium, a smaller number of domain walls is required and this indicates the increase of the period of the domain structure *d* with the increase of the degree of screening. At the same time, when screening with free charge carriers, starting at a certain concentration of carriers *n* the equilibrium width *d* abruptly increase to infinity, i.e. a monodomain structure is formed in the crystal.

The above is very well illustrated with the help of energy diagrams in Fig.1.5 representing the surface density of the energy of the depolarizing field, the domain walls and their sum in crystals with different degree of screening. Comparison of the diagrams *a*, *b* and *c* in this graph shows that the dependence of the energy of the depolarizing field on *d* in the presence of screening is no longer described by a straight line and has the form of a more complicated curve *1*. Its origin at small *d* coincides with the corresponding straight line without screening, and at high *d* reaches the asymptotic value describing the energy of the depolarizing field in the presence of

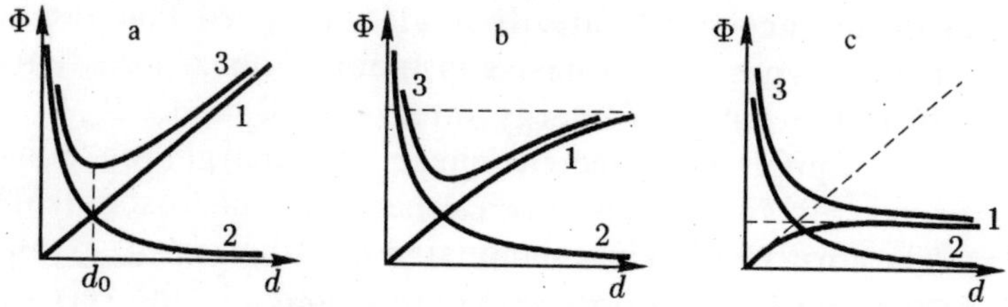


**Fig. 1.5.** Dependence on the average width of the domain *d* of the surface density of the depolarizing field (1), surface density of the energy of domain walls (2) and the sum of these energies for the following cases: (a) – no screening, (b) – weak screening, (c) – strong screening.

screening in the monodomain crystal. As the result the sum of curves *1* and *2*, i.e. curve *3* changes, the minimum of which corresponds to the equilibrium width of domain $d$.

Graph *b* in Fig. 1.5 shows that the point of intersection of curves *1* and *2* with screening taken into account is shifted to the right in comparison with the point of intersection of these curves without screening. Thus, the presence of even weak screening increases the period of the domain structure. For relatively strong screening (graph *c*, Fig. 1.5) starting with the case when curve *2* intersects curve *1* in the area where it reaches saturation, curve 3 does not have a minimum at all at finite values of $d$. The minimum value $\Phi$ is realized here at $d \to \infty$, which corresponds to transition to the monodomain state. According to the above considerations, in order to evaluate the critical concentration of carriers resulting in monodomain formation, it is necessary to equate simply the Debye screening length on which the field drops in the presence of screening, to the equilibrium width of the domain $d$ determined by equation (2.1). As shown later, this leads precisely to the equation for the critical concentration of the carriers obtained from more accurate estimates.

For more detailed description of these phenomena, let us consider initially only the influence of charges on surface states. For a straight cut plate, the influence of the charges located on the surface levels on screening of polarization, formed at phase transition is taken into account by means of the appearance of an additional term in the condition of induction continuity at the boundary of the surface layer (2.4). When writing down equation (2.4), it is necessary to specify a model of surface states. Let's assume that on the external surface of the investigated material there are both donor and acceptor states with the surface concentrations $N_d$ and $N_a$, respectively, and ionisation of the donor centre on the surface is accompanied by capture of the released electron on the acceptor state. It is also assumed that the surface states of both types form quasi-continuous zones, i.e. distributed uniformly in the range of energy intervals $\Delta E_d$ and $\Delta E_a$.

In the non-polar paraelectric phase, the charges on donor and acceptor centres compensate each other both macroscopically and locally. The formation of the modulated distribution of spontaneous polarization and the appropriate bound charges in the ferroelectric phase on the surface of the ferroelectric results in the redistribution of charges on the surface states, so that in areas with the positive potential there appears a large number of negatively charged acceptor centres, and vice versa: in areas with the negative potential there appears a larger number of positively charged donor centres, that have lost electrons.

To write down the boundary condition (2.4), we determine in advance the surface density of the charge on the surface states and its relation to the potential $\varphi$. To be more precise, it is assumed that $\Delta E_a = \Delta E_d \equiv \Delta E$ and $N_a = N_d \equiv N_s$. In addition to this, the energy ranges of the distribution of donor and acceptor centres overlap so both kinds of states are present both above and below the Fermi level. In this case, when the bound charge is formed on the surface a charge proportional to $\varphi$ and equal to $N_s e^2 \varphi \Delta E$ is carried over to the area with the positive potential and the charge of the same value is released on the donor centres at the same time. Consequently, the total surface density of the charge on the surface states is equal to

$$\frac{2N_s e^2 \varphi}{\Delta E}\Big|_{z=L/2+\Delta}. \tag{4.1}$$

Taking this into account, the first of the boundary conditions (2.4) is written down in the form

$$\varepsilon E_{II} - E_I = \frac{8\pi N_s e^2}{\Delta E}\varphi\Big|_{z=L/2+\Delta} = \frac{1}{\Lambda}\varphi\Big|_{z=L/2+\Delta}, \tag{4.2}$$

and the other equations, forming the set determining the dependence $\alpha_z(k)$, remain unchanged. The study of this system taking into account the change of (4.2) yields the following equation determining the dependence $\alpha_z(k)$ in the case of polarization screening by charges on the surface states:

$$\operatorname{tg} k\sqrt{\frac{\varepsilon_x}{\varepsilon_z}}.\frac{L}{2} = \frac{\sqrt{\varepsilon_x \varepsilon_z}}{\varepsilon}\frac{[(\varepsilon+(1+1/k\Lambda))+(\varepsilon-(1+1/k\Lambda))e^{-2k\Delta}]}{[(\varepsilon+(1+1/k\Lambda))-(\varepsilon-(1+1/k\Lambda))e^{-2k\Delta}]}. \tag{4.3}$$

In the absence of screening, i.e. at $\Lambda \to \infty$, equation (4.3) changes naturally to the already known relationship (2.5). In the absence of the surface layer but in the presence of screening, i.e. when the surface states are located directly on the surface of ferroelectric material ($\Delta = 0$) the dependence $\alpha_z(k)$ is determined by the condition

$$\operatorname{tg} k\sqrt{\frac{\varepsilon_x}{\varepsilon_z}}.\frac{L}{2} = \frac{\sqrt{\varepsilon_x \varepsilon_z}}{(1+1/k\Lambda)}. \tag{4.4}$$

In the presence of free surface carriers in the volume of the specimen simultaneously with charges on the surface states in the previous consideration the potential $\varphi_{III}$ in the volume of the material should be replaced by the potential

$$\varphi_{\mathrm{III}} = D\sin\left(\sqrt{\frac{k^2\varepsilon_x + (1/\lambda^2)}{\varepsilon_z}}\,z\right) \tag{4.5}$$

with the Debye screening length

$$\lambda = \sqrt{\frac{kT}{4\pi e^2 n_0}} \tag{4.6}$$

in the case when the crystal contains a dopant of mainly one type with the concentration of ionised centres equal to $n_0$. In this case the condition, determining the dependence of $\alpha_z$ on the wave vector $k$ for $\Delta = 0$ is rewritten as follows:

$$\frac{\sqrt{[k^2\varepsilon_x + (1/\lambda^2)]\varepsilon_z}}{k}\,\mathrm{ctg}\left(\frac{\sqrt{[k^2\varepsilon_x + (1/\lambda^2)]}}{\varepsilon_z}\frac{L}{2}\right) = 1 + \frac{1}{k\Lambda}. \tag{4.7}$$

At $\Lambda$, $\lambda \neq 0$ the dependence $\alpha_z(k)$ in this case is determined by the equation

$$\alpha_z(k) = \alpha_0(T - T_c) + \varkappa k^2 + \frac{\pi^3[\varepsilon_x + (1/k^2\lambda^2)]}{(\varepsilon_x kL/2 + L/2k\lambda^2 + 1/k\Lambda)^2}. \tag{4.8}$$

At $\Lambda \to \infty$, this equation transforms into the relation

$$\alpha_z(k) = \alpha_0(T - T_c) + \varkappa k^2 + \frac{\pi^3}{k^2\lambda^2\varepsilon_x + 1}\cdot\frac{4\lambda^2}{L^2}. \tag{4.9}$$

As a result of stability loss, the system will transform to the state with the wave vector $k$ corresponding to the condition $\partial\alpha_z/\partial k = 0$. In the presence of screening by only free charges in the bulk of the crystal, according to (4.9) the corresponding value of $k$ is determined by the expression [44]

$$k^2 = \frac{2\pi\sqrt{\pi}}{\sqrt{\varepsilon_x \varkappa L}} - \frac{1}{\varepsilon_x\lambda^2}. \tag{4.10}$$

Equation (4.10) shows clearly that at

$$\lambda^2 = \frac{\sqrt{\varkappa L}}{2\pi\sqrt{\pi\varepsilon_x}} \tag{4.11}$$

i.e. at

$$n_0 = \frac{\sqrt{\pi}}{2}\frac{kT\sqrt{\varepsilon_x}}{e^2\sqrt{\varkappa L}} \tag{4.12}$$

the period of the onsetting structure tends to infinity, which corresponds to transition to the monodomain state.

The estimates of critical concentration $n_0$ from (4.12) at $T \sim 300$ K, $\varkappa \sim a^2 \sim 10^{-14}$ cm$^2$, $L \sim 10^{-1}$ cm yield the value of $n_0 \sim 10^{13}$cm$^{-3}$. The corresponding shift of $T_c$ in comparison with the infinite crystal is in this case equals to:

$$\Delta T_c = \frac{4\pi^3\lambda^2}{\alpha_0 L^2}. \tag{4.13}$$

For the found concentration $n_0$ this value of $\Delta T_c$ is estimated at $\Delta T_c \sim 10^{-2}$ K.

On the other hand, at finite $\Lambda$ and $\lambda \to \infty$, instead of (4.8) we have the following dependence

$$\alpha_z(k) = \alpha_0(T - T_c) + \varkappa k^2 + \frac{\pi^3 \varepsilon_x}{(\varepsilon_x kL/2 + 1/k\Lambda)^2}. \tag{4.14}$$

It differs from (2.8) in the following: electrostatic contribution in $\alpha_z(k)$ is no longer a monotonically dropping function $k$, but passes through a maximum and tends to zero due to the efficiency of screening in equilibrium at low $k$. Consequently, the overall dependence $\alpha_z(k)$ in the general case will have absolute maximum at $k = 0$ and under certain ratio of the parameters it will have a local minimum at $k \neq 0$. The extrema of this dependence are determined by the equation

$$k\left[\frac{\varkappa}{\pi^3 \varepsilon_x \Lambda^2} - \frac{\left(\tilde{k}^2 - 1\right)}{\left(\tilde{k}^2 + 1\right)}\right] = 0, \quad \tilde{k}^2 = \frac{\varepsilon_x L \Lambda}{2} k^2. \tag{4.15}$$

Equation (4.15) shows that the local maximum in the dependence $\alpha_z(k)$ will be observed at

$$k_1 = \frac{4\sqrt{\varkappa}}{\pi\sqrt{\pi \varepsilon_x \Lambda}\sqrt{\Lambda L}}, \tag{4.16}$$

and the local minimum in the first approximation at

$$k_2 = \frac{\sqrt{2}\pi^{3/4}}{\varepsilon_x^{1/4} \varkappa^{1/4} L^{1/2}}. \tag{4.17}$$

The considered local state becomes unstable at $k_1 = k_2$, i.e. at

$$\Lambda = \frac{2\sqrt{\varkappa}}{\pi\sqrt{\pi}\sqrt{\varepsilon_x}}. \tag{4.18}$$

Taking into account the fact that according to the order of magnitude $\varkappa \sim a^2$, where $a$ is the size of the elementary cell, equation (4.15) shows clearly that in this case $\Lambda < a$. In accordance with the definition this takes place at $N_s \sim 10^{14}\text{cm}^{-2}$, i.e. at the maximum possible density of the surface electronic states.

It should be mentioned that within the framework of the proposed model, surface screening is linked with the migration of charges along the surface over the distance of the order of the wavelength of the onsetting phase. Evidently, in the conditions of real cooling of the specimen with a finite rate, the migration of the charges over large distances and, therefore, the efficiency of screening at low $k$ are impeded. As the result the state corresponding to the absolute minimum of the thermodynamic potential will most probably be not implicated and the state corresponding to the local minimum will take place (Figs. 1.6 and 1.7). In the presence of a finite but not very strong screening, this state corresponds to the half period of

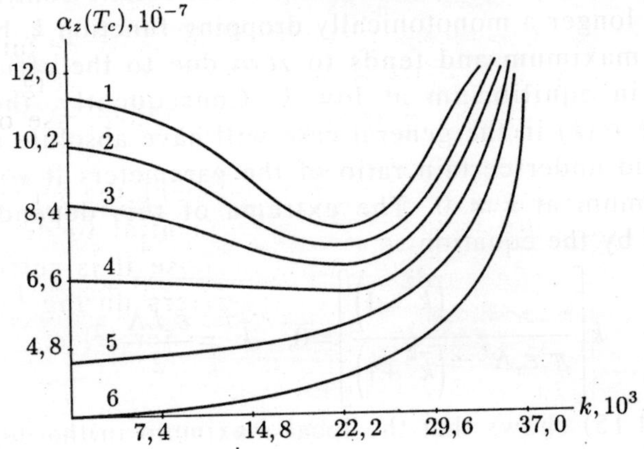


**Fig. 1.6.** Behaviour of the dependence $\alpha_z(k)$ in the vicinity of the local minimum for ferroelectrics with charges on surface states. 1, 2, 3, 4, 5, 6 – $\Lambda^{-1} = 1\cdot10^8$; $1.2\cdot10^8$; $1.5\cdot10^8$; $2\cdot10^8$; $3\cdot10^8$; $4\cdot10^8$; $\Delta = 0$.

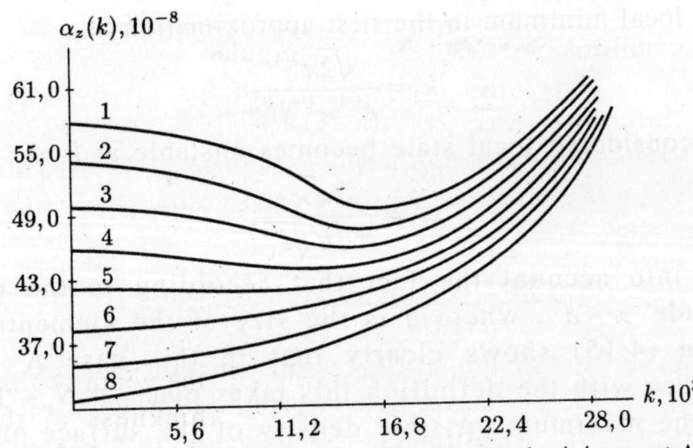


**Fig. 1.7.** Dependence of $\alpha_z(k)$ in the vicinity of the local minimum at $\Lambda^{-1} = 4\cdot10^8$ and various $\Delta$: 1,2,3,4,5,6,7,8 – $\Delta$ = $1.5\cdot10^{-8}$; $1.3\cdot10^{-8}$; $1\cdot10^{-8}$; $8\cdot10^{-9}$; $6\cdot10^{-9}$; $6\cdot10^{-9}$; $4\cdot10^{-9}$; 0.0.

the heterogenous distribution of polarization

$$d = \pi \bigg/ \left( \frac{2\pi\sqrt{\pi}}{\sqrt{\varkappa \varepsilon_x} \cdot L} - \frac{32\pi N_s e^2}{\Delta E L \varepsilon_x} \right)^{1/2}, \tag{4.19}$$

which increases with the increase of the density of surface states.

Thus, surface screening also demonstrates a tendency for the increase of the size of the domain structure. In the framework of the model under consideration this tendency should be restricted to the case of at least two domains in the crystal on the basis of the condition of equality to zero of the total charge on the surface states on each of crystal surfaces that are perpendicular to the vector of spontaneous polarization.

Analysis of the dependence $\alpha_z(k)$ on the basis of the initial ratio (4.5) at various $\Delta$ and the fixed value of $\Lambda$ shows (Fig. 1.7) that the qualitative decrease of $\Delta$ is similar to the decrease of $\Lambda$, i.e. to the increase of $N_s$.

It should be mentioned that to implement the monodomain formation conditions (4.11), it is not essential to deal with a ferroelectric-semiconductor. For this purpose it is sufficient to create the required concentration of carriers during the phase transformation (for example by illuminating a ferroelectric material by the light of required frequency). Surface screening may also be created purposefully, by forming a special structure of defective centres on the surface.

## 1.5. Formation of the domain structure during inhomogeneous cooling of ferroelectrics

The real conditions of transition to the polar state usually imply the presence of a temperature gradient in a specimen being rapidly cooled which, as shown below, has a significant effect on the period of the resultant structure.

The result of the influence of inhomogeneous cooling on the domain structure may easily be predicted if it is noted that in a inhomogeneously cooled specimen the volume of the part of the material, undergoing phase transition at the moment of nucleation of the domain structure, decreases. From the viewpoint of calculations, this means that while estimating the width of the domain the equation (1.14) should include the thickness of the layer undergoing phase transition and not the thickness of the specimen $L$ (Fig. 1.8). Since the former is evidently smaller than the thickness of the specimen and decreases with increasing temperature gradient,

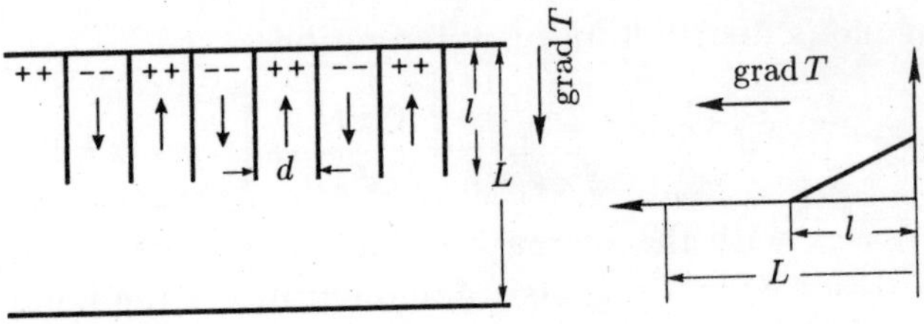


**Fig. 1.8.** Formation of a domain structure in a ferroelectric plate in the conditions of the temperature gradient.

in accordance with the mentioned equation it should result in the decrease of the width, i.e. in the refining of the domain structure with the increase of the cooling rate.

Let us initially consider the case of a 'pure' ferroelectric while making quantitative calculations. It is assumed that its surface, perpendicular to the ferroelectric axis, coincides with plane $z = 0$, and the bulk of the crystal corresponds to values of $z > 0$. Let us study the conditions of formation and characteristics of the plane-parallel domain structure which is periodic along the $x$ axis. It is assumed that the free surface of the ferroelectric crystal is cooled down below the Curie temperature $T_c$ and the remaining volume of the crystal is in the paraphase generated by the temperature gradient $\partial T/\partial z$ and directed into the volume of the ferroelectric crystal normally to the surface.

As in section 1.2, the distribution of the electric fields in the vicinity of the surface of the ferroelectric crystal is determined by the electrostatic equation (2.2) where

$$\varepsilon_x = 1 + \frac{4\pi}{\alpha_x}, \quad \varepsilon_z = 1 + \frac{4\pi}{-\alpha_z + \alpha_1 z + \varkappa k^2}, \qquad (5.1)$$
$$\alpha_z = \alpha_0 (T - T_c), \quad \alpha_1 = \frac{\partial \alpha}{\partial z}.$$

For the periodic distribution of the potential $\varphi$ along axis $x$ with wave vector $k$, equation (2.2) taking (5.1) into account is converted to the following form

$$-\varepsilon_x k^2 \varphi + \frac{\partial}{\partial z}\left( \frac{4\pi}{-\alpha_z + \alpha_1 z + \varkappa k^2} \right) \frac{\partial \varphi}{\partial z} = 0 \qquad (5.2)$$

which takes into account the explicit dependence of dielectric permittivity $\varepsilon_z$ on coordinate $z$.

By conversion to dimensionless quantities

$$k_1 = k\frac{\varkappa^{3/8}}{\alpha_1^{1/4}}, \quad z_1 = z\frac{\alpha_1^{1/2}}{\varkappa^{1/4}}, \quad z_3 = \left(\frac{\varepsilon_x k_1}{4\pi}\right)^{1/3} z_2,$$

$$z_2 = -\frac{\alpha_z}{\alpha_1^{1/2}\varkappa^{1/4}} + z_1 + k_1^2 \tag{5.3}$$

equation (5.2) is reduced to the form

$$\frac{\partial^2\varphi}{\partial z_3^2} - \frac{1}{z_3}\frac{\partial\varphi}{\partial z_3} - z_3\varphi = 0. \tag{5.4}$$

Its solution has the form

$$\varphi(z_3) = z_3 Z_{2/3}\left(\frac{2}{3}z_3^{3/2}\right), \tag{5.5}$$

where $Z_{2/3}(x)$ is any solution of the Bessell equation of the order of 2/3.

Taking into account the fact that the value of the potential at infinity should convert to zero, we select the Bessell function with the corresponding asymptotic. Consequently, solution (5.5) of equation (5.4) becomes more specific as shown below:

$$\varphi(z_3) = z_3 K_{2/3}\left(\frac{2}{3}z_3^{3/2}\right) \simeq \sqrt{\frac{\pi z_3}{2}}\exp\left(-\frac{2}{3}z_3^{3/2}\right). \tag{5.6}$$

The equation for determination of the domain structure parameters is found from the condition of equality to zero of the field on the cooled surface

$$\frac{\partial\varphi}{\partial z}\bigg|_{z=0} = 0. \tag{5.7}$$

Substituting (5.6) into (5.7) and taking into account (5.3) we obtain the dependence

$$\alpha_z(k) = \varkappa k^2 + \left(\frac{\pi\alpha_1}{\varepsilon_x k^2}\right)^{1/3}. \tag{5.8}$$

The transition to the polar phase takes place in the state corresponding to the condition $\partial\alpha_z/\partial k = 0$, i.e. in the state with

$$k_m = \left(\frac{\pi\alpha_1^2}{\varepsilon_x\varkappa}\right)^{1/8}. \tag{5.9}$$

Equation (5.9) shows that the appearing structure is refined with an increase of the temperature gradient. The result is completely clear because in the presence of the gradient the transition is observed not in the entire volume of the material but only in the layer of the material which, according to (2.9), should lead to (instead of $L$ – the thickness of the layer in which the transformation takes place) reduction of the resultant structure.

The described structure forms under the condition in which supercooling on the surface of the ferroelectric specimen in comparison with a infinite material reaches the value of the following order

$$\Delta T \sim \left(\frac{\varkappa \alpha_1^2}{\varepsilon_x}\right)^{1/4} \frac{1}{\alpha_0}, \tag{5.10}$$

which at used above values of $\varkappa$, $\varepsilon_x$, and the value of $\alpha_1$, corresponding to the temperature gradient in the specimen of the order of 10 K/cm, $\alpha_0 \sim 10^{-3}$ $K^{-1}$ equals to $\sim 10^{-1}$ K.

Evidently, the inhomogeneous distribution of temperature in the specimen is observed during rapid cooling of the latter. In this instance in accordance with the results of this section a structure with a smaller period is actually observed [50] as compared to slow cooling.

## 1.6. Formation of the domain structure in ferroelastic contacting a substrate, and in material with a free surface

The formation of a domain structure in a ferroelastic material that contacts a substrate that does not undergo phase transition, as in the case of ferroelectrics, is associated with a decrease of the energy of long-range elastic fields formed in the vicinity of contact both in the case of contact with an elastic and with the absolutely rigid substrate [39, 41].

This will be illustrated by the example of contact of the ferroelastic with an absolutely rigid substrate leading to clamping of the ferroelastic material in the contact zone. Let the ferroelastic have the form of a plate with thickness $L$ with the normal to the surface coinciding with axis $z$, and the displacements in the material in process of phase transition $u$ coincide with the axis $y$.

The thermodynamic potential of the ferroelastic is as follows

$$\Phi = \int \left[\frac{\alpha}{2}\left(\frac{\partial u}{\partial x}\right)^2 + \frac{\varkappa}{2}\left(\frac{\partial^2 u}{\partial x^2}\right) + \frac{c}{2}\left(\frac{\partial u}{\partial z}\right)^2\right] dx\,dz, \tag{6.1}$$

where the critical modulus $\alpha = \alpha_0(T-T_c)$, $c$ is the optional elastic modulus, $\varkappa$ is a correlation parameter. When writing down (6.1), the gradient member is left only on axis $x$, i.e. it is assumed that for the other directions the correlation effects are small. Equation (6.1) is written down under condition that only small vicinity of $T_c$ is examined where the nonlinearity may be ignored because of the small strain amplitude.

Stress tensor components that differ from zero are found as derivatives $\sigma_{ik}=\partial\Phi/\partial u_{ik}$ of (6.1), which yields

$$\sigma_{12}=\alpha\frac{\partial u}{\partial x}-\varkappa\frac{\partial^2}{\partial x^2}\frac{\partial u}{\partial x},$$
$$\sigma_{23}=c\frac{\partial u}{\partial z}. \tag{6.2}$$

For periodic distribution of displacements along axis $x$ with the wave vector $k$ from the equation of elastic equilibrium

$$\nabla_i\sigma_{ik}=0 \tag{6.3}$$

we discover the equation for displacement $u$ of the following type

$$\left(\alpha-\varkappa k^2\right)k^2u+c\frac{\partial^2 u}{\partial z^2}=0. \tag{6.4}$$

The solution of equation (6.4) has to meet specific conditions at the boundary of the material. In the present case they can be represented by

$$u\big|_{z=0}=0,\quad \frac{\partial u}{\partial z}\Big|_{z=L}=0, \tag{6.5}$$

i.e. it is assumed that at $z = 0$ there is a contact with the absolutely rigid material, and the second boundary of the material $z = L$ is assumed to be free. The following function meets equation (6.4) and conditions (6.5):

$$u=B\sin\frac{\pi}{2L}z. \tag{6.6}$$

Substituting (6.6) into (6.5) we obtain the dependence

$$\alpha=\varkappa k^2+c\left(\frac{\pi}{2kL}\right)^2. \tag{6.7}$$

The value of $k$ corresponding to the minimum of dependence $\alpha(k)$

$$k_m=\left(\frac{c\pi^2}{4\varkappa L}\right)^{1/4}. \tag{6.8}$$

At usual $c\sim 10^{10}$ erg·cm$^{-3}$, $\varkappa\sim c\cdot a^2$, $a^2\sim 10^{-15}$ cm$^2$, $L\sim 10^{-1}$ cm, the period of the resultant structure $d=\pi/k_m$ has the order of $10^{-4}$ cm which corresponds to the experimentally observed domain dimensions [2, 12, 16]. The shift of the phase transition temperature in relation to $T_c$ of an infinite crystal is

$$\Delta T=\frac{\pi(\varkappa c)^{1/2}}{\alpha_0 L}. \tag{6.9}$$

In the absence of the substrate, i.e. in the ferroelastic material with a free surface, in the case under consideration there are no elastic fields in the vicinity of the surface whatsoever and the very fact of formation of the regular domain structure becomes difficult to understand. Evidently, in this case the structure is metastable because of the uncompensated positive energy of the domain boundaries.

As shown in [43], the situation with the presence of the domain structure in an unclamped ferroelastic may be understood if it is taken into account that in the real conditions the phase transformation usually takes place in the presence of a temperature gradient in the specimen due to its inhomogeneous cooling. Under these conditions, as a result of its temperature dependence, spontaneous deformation changes along the direction of the temperature gradient, i.e. along the normal to the surface of the specimen. This heterogeneity of deformation similarly to the case of contact with the substrate results in the formation of elastic fields. To reduce the energy of these fields, the specimen transfers to the state with the distribution of the deformation modulated along the surface (Fig. 1.9).

Let us consider this transition more thoroughly. The distribution of displacements and of the accompanying elastic stresses in the ferroelastic material, as in the case of ferroelectric materials, is described by simultaneous equations in which the role of material equations is performed by the conventional Hooke law with additional terms corresponding to the correlation effects (6.2) and the role of Laplace's equation is performed by the elastic equilibrium equation (6.3). As previously, the displacements in the material formed in the process of spontaneous deformation are assumed to be coincident with axis $y$. The presence of the temperature gradient is taken into account

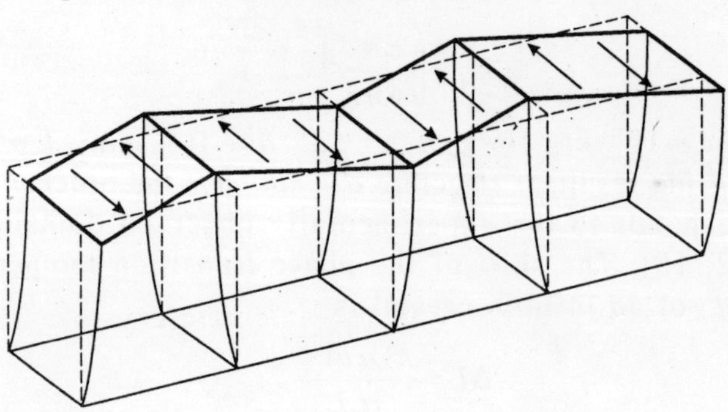


**Fig. 1.9.** Distribution of displacements in a ferroelastic material contacting with an absolutely rigid substrate ($z = 0$).

by the coordinate dependence of the critical modulus $\alpha$ in (6.2):

$$\alpha(z)=\alpha_0(T-T_c)+\frac{\partial\alpha}{\partial z}\cdot z. \tag{6.10}$$

Taking this into account equation (6.4) for displacements with the help of dimensionless quantities:

$$z_3=k_1^{2/3}(z_1-k_1+\alpha_{02}),\quad z_1=z\frac{\alpha_1^{1/2}}{\varkappa_1^{1/4}},\quad \alpha_1=\frac{\alpha'}{c},$$

$$\alpha'=\partial\alpha/\partial z,\quad \varkappa_1=\varkappa/c,\quad k_1=k\frac{\varkappa_1^{3/8}}{\alpha_1^{1/4}}, \tag{6.11}$$

$$\alpha_{02}=\frac{\alpha_{01}}{\alpha_1^{1/2}\varkappa^{1/4}},\quad \alpha_{01}=\alpha_1/c$$

leads to the equation for the Airey function

$$\frac{\partial^2 u}{\partial z_3^2}-z_3u=0. \tag{6.12}$$

Its solution should satisfy the boundary conditions:

$$u\big|_{z=\infty}=0,\quad \frac{\partial u}{\partial z}\Big|_{z=0}=0, \tag{6.13}$$

which gives the dependence of the modified coefficient α on the wave vector $k$ on the cooled surface of the following type:

$$\alpha(k)=\alpha_0(T-T_c)+\left(\frac{c\alpha'^2}{k^2}\right)^{1/3}+\varkappa k^2 \tag{6.14}$$

From minimisation of $\alpha$ in respect of $k$ we find the value of $k$ that determines the period of the structure condensed at phase transition

$$k_m=\left(\frac{c\alpha'^2}{27\varkappa^3}\right)^{1/8}. \tag{6.15}$$

This structure is implemented under the condition when supercooling $\Delta T$ on the surface of the ferroelastic specimen reaches the value of the order

$$\Delta T=\frac{\varkappa^{1/4}\alpha'^{1/2}c^{1/4}}{\alpha_0}. \tag{6.16}$$

## 1.7. The fine-domain structure in ferroelectric crystals with defects

A domain structure, repeating in a specific manner the distribution of defects, may be formed in the vicinity of the Curie point in ferroactive crystals with defects. For 'strong' defects, the minimum size of such a domain is close to the average distance between the defects, for 'weak' defects it is equal to the size of the area within which a sufficiently strong fluctuation of the defect concentration occurs. In any case, the domain size is usually considerably smaller here than the mean equilibrium width of the domains in the perfect crystal and, therefore, the domain structure in defective materials is referred to as the fine-domain structure.

The formation of the fine-domain structure is associated with the so called 'polar' defects. In the vicinity of the Curie point at which the crystal structure of the ferroelectric is extremely susceptible to external effects, these defects polarize the lattice and create in the crystal a specific distribution of polarization replicating the distribution of electric fields of defects $E_d$. Due to the chaotic orientation of the polar defects in this polarization distribution there are evidently areas with both positive and negative polarization. With the decrease of temperature while the temperature moves away from the Curie point the initially relatively smooth distribution of polarization from point to point is replaced by an almost step-like distribution with relatively homogeneous polarization within the limits of each domain and distinct domain boundaries.

This takes place at such temperatures when the width of the domain wall becomes considerably smaller than the mean distance between the defects. Since this moment, we may consider the formation of a domain structure is created by defects. The further decrease of temperature results in a comparatively rapid increase of the energy of domain boundaries $\sim(\Delta/T)^{3/2}$ ($\Delta T$ is the distance from Curie point $T_c$) in comparison with the temperature dependence of gain in the volume energy $\sim E_d P_0 \sim (\Delta/T)^{1/2}$, which yields domain formation on a defect. If these energies are equal

$$\gamma = 2P_0 E_d d, \tag{7.1}$$

the considered domain structure losses its stability. Equation (6.10) makes it possible to estimate the temperature range in which the fine-domain structure exists linked with defects. As expected, this range is small and is equal to several degrees.

Completing this chapter we can name a whole series of factors influencing the parameters of the domain structure. These are the dimensions and dielectric permittivity of the surface non-ferroelectric layer, the type of cut used in preparation of the specimen, electrodes, and presence of the volume or surface screening by charge carriers during phase transformation, crystal structure defects, transition under the condition of a temperature gradient for ferroelectrics, the type of substrate with special dimensions and elastic properties, and inhomogeneous cooling for the ferroelastics. Varying these parameters, it is possible to produce the required type of domain structure.

Chapter 2

# Structure of domain and interphase boundaries in defect-free ferroelectrics and ferroelastics

## 2.1. Structure of 180° domain boundary in ferroelectrics within the framework of continuous approximation in crystals with phase transitions of the first and second order

The formed domain structure is characterized by distinctive boundaries between the domains the so-called domain walls, within which the entire variation of polarization or deformation from the values corresponding to one domain to the values corresponding to the adjacent domain are concentrated. The width of the domain wall is usually considerably smaller than width $d$ of the domain itself. When considering the structure of the domain wall, we can ignore the effect of other boundaries and investigate an isolated domain wall. The possible effect on the domain wall structure of the depolarizing field of bound charges on the surface of a ferroelectric, which will be considered in section 2.3, should in any case be restricted by the thickness of the layer within which the given field penetrates into the material. As it can be seen from formula (1.6) in chapter 1 in particular, the thickness of this layer has the order of the width of the domain. At a large distance from the surface of the ferroelectric inside the bulk of the material the influence of these fields can be ignored and in investigation of the structure of the wall we can use the approximation of the infinite material.

Taking these restrictions into account, let us consider the simplest case of 180° domain wall in a ferroelectric crystal with a phase transition of the second order. Let us use here the so-called continuous approximation, which ignores the discreteness of the crystal lattice.

Let, as previously, the polar direction in the crystal coincide with the axis $z$, and the plane of the domain wall with the plane $zy$, so that the distribution of polarization in the transition layer between the domains depends only on the distance along the normal to the plane of the wall: $\mathbf{P} = \mathbf{P}(x)$.

The majority of ferroelectrics are characterized by a very high energy of anisotropy so that the structure of the wall of the rotating type, identical to ferromagnetics, is unfavorable [18]. In such domain walls, the polarization vector without changing its length at every point of the boundary rotates through 180° within the limits of the boundary (possible cases of the formation of rotating boundaries in ferroelectrics will be discussed in section 2.6). At a high anisotropy energy, the spatial variation of polarization vector $\mathbf{P}$ in the boundary is linked with the variation of its modulus $|\mathbf{P}| = P_z \equiv P(x)$.

In the framework of continuous approximation, the structure of the wall is determined by the minimum of the thermodynamic potential, in which to the usual local contribution of $\varphi(P) = -\frac{\alpha}{2}P^2 + \frac{\beta}{4}P^4$, $-\alpha \equiv \alpha_z = \alpha_0(T-T_c)$, (it is sufficient under consideration of a homogeneous material) we add the so-called correlation term $\frac{\varkappa_{ilkm}}{2}\frac{\partial P_i}{\partial x_k}\frac{\partial P_l}{\partial x_m}$, where $\varkappa_{ilkm}$ is the tensor of correlation constants. In the present case $\mathbf{P} \equiv P_z$, $P = P(x)$ and from the entire set we retain here only one correlation term with the constant $\varkappa_{3311} \equiv \varkappa$ equal to $\frac{\varkappa}{2}\left(\frac{dP}{dx}\right)^2$. Since the density of the thermodynamic potential $\overline{\Phi} = \varphi(P) + \frac{\varkappa}{2}\left(\frac{dP}{dx}\right)^2$ changes within the limit of the boundary from point to point, the structure of the wall in this case is determined by the minimum of the functional $\Phi = \int \overline{\Phi}\, dx$:

$$\Phi = \int_{-\infty}^{\infty}\left\{-\frac{\alpha}{2}P^2 + \frac{\beta}{4}P^4 + \frac{\varkappa}{2}\left(\frac{dP}{dx}\right)^2\right\}dx. \tag{1.1}$$

To determine the optimum distribution $P(x)$, corresponding to the minimum $\Phi$, let us vary $\Phi$ in respect of $P$. For this purpose we write in advance

$$\varphi(P+\delta P) = \varphi(P) + \frac{\varphi'(P)}{1!}\delta P + \frac{\varphi''(P)}{2!}(\delta P)^2, \tag{1.2}$$

$$\frac{\varkappa}{2}\left[\frac{d}{dx}(P+\delta P)\right]^2=\frac{\varkappa}{2}\left[\frac{dP}{dx}+\frac{d(\delta P)}{dx}\right]^2=$$
$$=\frac{\varkappa}{2}\left(\frac{dP}{dx}\right)^2+\varkappa\frac{dP}{dx}\frac{d(\delta P)}{dx}+\frac{\varkappa}{2}\left[\frac{d(\delta P)}{dx}\right]^2. \tag{1.3}$$

Then

$$\Phi(P+\delta P)=\int\left\{\varphi(P+\delta P)+\frac{\varkappa}{2}\left[\frac{d(P+\delta P)}{dx}\right]^2\right\}dx=$$
$$=\int\left\{\varphi(P)+\frac{\varkappa}{2}\left(\frac{dP}{dx}\right)^2\right\}dx+$$
$$+\int\left\{\frac{\varphi'(P)}{1!}\delta P+\varkappa\frac{dP}{dx}\frac{d(\delta P)}{dx}\right\}dx+ \tag{1.4}$$
$$+\int\left\{\frac{\varphi''(P)}{2!}(\delta P)^2+\frac{\varkappa}{2}\left[\frac{d(\delta P)}{dx}\right]^2\right\}dx\equiv\Phi(P)+\delta\Phi+\delta^2\Phi.$$

The first term in the right-hand part of (1.4) describes the thermodynamic potential of the optimum distribution $P(x)$, with respect to which the variation is performed. It coincides with expression (1.1). The following terms represent respectively the first and second variations of the potential (1.1).

The equality to zero of the first variation $\delta\Phi = 0$ enables us to find the distribution $P(x)$, corresponding to the minimum $\Phi$. Taking into account integration by parts

$$\int\left\{\varphi'(P)\delta P+\varkappa\frac{dP}{dx}\frac{d(\delta P)}{dx}\right\}dx=\int\left\{\varphi'(P)-\varkappa\frac{d^2P}{dx^2}\right\}\delta P dx. \tag{1.5}$$

Since the variation $\delta P$ is an arbitrary small function, the identical equality to zero of the integral is possible only if the expression in the braces is equal to zero. From this we find the equation describing distribution of polarization in the boundary:

$$\varkappa\frac{d^2P}{dx^2}=\frac{d\varphi}{dP}=-\alpha P+\beta P^3. \tag{1.6}$$

The sign of the second variation makes it possible to evaluate the stability of the corresponding solution. Similarly as in (1.5)

$$\delta^2\Phi = \int\left\{\frac{\varphi''(P)}{2!}(\delta P)^2 + \frac{\varkappa}{2}\left[\frac{d(\delta P)}{dx}\right]^2\right\}dx =$$

$$= \int\frac{1}{2}\left\{\frac{d^2\varphi}{dP^2}\delta P - \varkappa\frac{d^2(\delta P)}{dx^2}\right\}\delta P dx = \tag{1.7}$$

$$= \int\delta P\left\{-\frac{\varkappa}{2}\frac{d^2}{dx^2}+\frac{1}{2}\frac{d^2\varphi}{dP^2}\right\}\delta P dx = \int\delta P\hat{L}\delta P dx,$$

where the differential operator is

$$\hat{L} = \left[-\frac{\varkappa}{2}\frac{d^2}{dx^2}+\frac{1}{2}\frac{d^2\varphi}{dP^2}\right]. \tag{1.8}$$

The problem of finding the sign of the second variation for determining the stability of the corresponding polarization distribution is reduced to investigation of the $\hat{L}$ operator spectrum. This spectrum, i.e. a set of eigenvalues $\lambda_n$ of the operator $\hat{L}$, can be found from the equation

$$\hat{L}\psi_n = \left[-\frac{\varkappa}{2}\frac{d^2}{dx^2}+\frac{1}{2}\left(-\alpha+3\beta P^2(x)\right)\right]\psi_n = \lambda_n\psi_n, \tag{1.9}$$

which has the form of a Schrödinger equation for a particle in a potential field

$$V(x) = \frac{1}{2}\left(-\alpha+3\beta P^2(x)\right). \tag{1.10}$$

In this case, the eigenvalue $\lambda_n$ and eigenfunctions $\psi_n$ play the role of the eigenvalues of the energy of the particle and its wave functions, respectively.

From the general theorems of quantum mechanics it is known that $\lambda_n$ is the increasing function of number $n$. Therefore, it turns out that in order to judge the stability of the corresponding solution, it is sufficient to find out the sign of the minimum eigenvalue $\lambda_0$ of the operator (1.8). Let us show this. The arbitrary variation $\delta P(x)$ can always be expanded into a series in respect of the eigenfunctions $\psi_n(x)$ of the operator $\hat{L}$:

$$\delta P(x) = \sum_n A_n\psi_n(x), \tag{1.11}$$

where $n$ is the number of eigenvalue and $A_n$ is the coefficient of expansion. Substituting into (1.7) the expansion (1.11) using the condition of orthonormalization of eigenfunctions $\psi_n(x)$:

$$\int \psi_n^*(x)\psi_m(x)dx = \delta_{nm}, \tag{1.12}$$

and taking also into account the determination of the eigenvalue of the operator $\hat{L}$ (1.9), we obtain

$$\delta^2\Phi = \sum_{n=0}^{\infty}\lambda_n |A_n|^2. \tag{1.13}$$

It is evident that in the presence of at least one eigenvalue $\lambda_n<0$, and the first negative value can only be the value $\lambda_0$, it is always possible to select such coefficients $A_n$, and, consequently, the required form of $\delta P$ that the second variation of the thermodynamic potential becomes negative. Thus, the condition of stability loss of some distribution $P(x)$ is the occurrence of the first negative eigenvalue $\lambda_0<0$ in the spectrum of the operator (1.8) [51].

Away from the boundary where the homogeneous state is implemented, the derivative $d^2P/dx^2 = 0$. The value of $P$ is determined here from the conventional equation

$$-\alpha P_0 + \beta P_0^3 = 0, \tag{1.14}$$

which in the case of the ferrophase gives $P_0^2 = \alpha/\beta$. According to (1.8) and (1.9) the condition of stability of such a state is trivial $\alpha > 0$.

The problems of investigation of the stability of polarization distribution that takes place in the domain boundary will be discussed in section 2.6.

To determine the structure of the 180° domain wall, we have to solve equation (1.6) using the following boundary conditions

$$\begin{cases} P(+\infty) = P_0 \\ P(-\infty) = -P_0. \end{cases} \tag{1.15}$$

To integrate the equation of the second order (1.6), let us find its first integral first of all. For this purpose, both parts of (1.6) are additionally multiplied by $dP/dx$ and integrated in respect of $dx$ taking into account the conditions (1.15). Consequently, the first integral of equation (1.6) has the form:

$$\frac{\varkappa}{2}\left(\frac{dP}{dx}\right)^2 = \varphi[P[x]] - \varphi[P_0]. \tag{1.16}$$

Separating the variables in (1.16) gives

$$\int \frac{dP}{\sqrt{\varphi[P(x)] - \varphi[P_0]}} = \int \frac{dx}{\sqrt{\varkappa/2}}. \tag{1.17}$$

Taking into account the specific values $P_0^2$, the difference $\varphi(P)-\varphi(P_0)$ is transformed to the form

$$\varphi[P(x)] - \varphi[P_0] = \frac{\beta}{4}\left[P_0^2 - P^2\right]^2. \tag{1.18}$$

Expanding the resultant difference of the squares into multipliers

$$\frac{1}{P_0^2 - P^2} = \frac{1}{2P_0(P_0+P)} + \frac{1}{2P_0(P_0-P)} \tag{1.19}$$

and integrating (1.17) taking (1.18) and (1.19) into account, we obtain

$$\frac{1}{\sqrt{\beta}P_0}\int\left(\frac{dP}{P_0+P} + \frac{dP}{P_0-P}\right) = $$

$$= \frac{1}{\sqrt{\beta}P_0}\ln\frac{P_0+P}{P_0-P} = \sqrt{\frac{2}{\varkappa}}(x-U). \tag{1.20}$$

The integration constant $U$ determines the position of the centre of the boundary where $P = 0$. Assuming in this case that the centre of the boundary is situated at the origin of the coordinates, i.e. at $U = 0$, and exponentiating ratio (1.20), we obtain the following distributing of polarization in the boundary [52]–[54]

$$P(x) = P_0 \cdot \text{th}\frac{x}{\delta}, \quad \delta = \frac{1}{P_0}\sqrt{\frac{2\varkappa}{\beta}} = \sqrt{\frac{2\varkappa}{\alpha}}. \tag{1.21}$$

In accordance with (1.21) and Fig. 2.1, quantity $\delta$ is naturally referred to as the half width of the domain boundary. As indicated

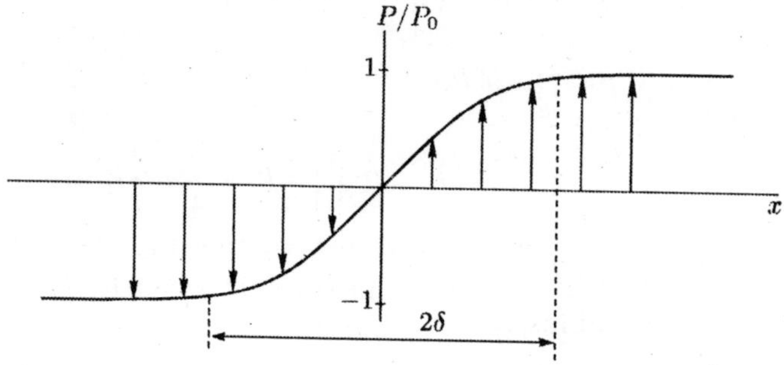


**Fig 2.1.** Distribution of polarization in the 180° boundary.

by (1.21), this quantity depends greatly on temperature and increases when approaching the Curie point $T_c$.

The surface density of the energy of the stationary wall $\gamma_0$ is obtained as a result of substituting the distribution (1.21) into (1.1) less the energy of the homogeneous state. Consequently, taking into account that $dP_0/dx = 0$, the first integral (1.16) and the ratio (1.18) we find

$$\begin{aligned}\gamma_0 &= \int_{-\infty}^{\infty}\left[\varphi(P)-\varphi(P_0)+\frac{\varkappa}{2}\left(\frac{dP}{dx}\right)^2\right]dx = \\ &= 2\int_{-\infty}^{\infty}\left[\varphi(P)-\varphi(P_0)\right]dx = \frac{\beta}{2}P_0^4\int_{-\infty}^{\infty}\left(1-\mathrm{th}^2\frac{x}{\delta}\right)^2 dx = \\ &= \frac{\alpha P_0^2\delta}{2}\int_{-\infty}^{\infty}\frac{dx/\delta}{\mathrm{ch}^4 x/\delta} = \frac{2}{3}\alpha P_0^2\delta = \frac{4}{3}P_0^2\frac{\varkappa}{\delta}.\end{aligned} \tag{1.22}$$

In a crystal with the phase transition of the first order, the expansion (1.1) contains the additional term $(\gamma/6)P^6$. Therefore, the distribution of polarization in the 180° boundary is described in this case by

$$\varkappa\frac{d^2P}{dx^2} = -\alpha P + \beta P^3 + \gamma P^5. \tag{1.23}$$

Away from the boundary in the homogeneous state in the ferrophase

$$P_0 = \frac{\beta}{\gamma}\left(\sqrt{1+\frac{\alpha\gamma}{\beta^2}}-1\right). \tag{1.24}$$

The first integration of equation (1.23) taking into account (1.24) and the boundary conditions $\left.\dfrac{dP}{dx}\right|_{\pm\infty} = 0$ yields

$$\begin{aligned}\frac{\varkappa}{2}\left(\frac{dP}{dx}\right)^2 &= \varphi\left[P(x)\right]-\varphi\left[P_0\right] = \\ &= \left(P_0^2-P^2\right)^2\left[\frac{\gamma}{6}P^2+\frac{\beta}{4}+\frac{\gamma}{3}P_0^2\right].\end{aligned} \tag{1.25}$$

Integration of (1.25) leads to the following distribution of polarization in the 180° boundary in the ferroelectric with a phase transition of the first order

$$P(x) = P_0 \frac{\text{sh}(x/\delta)}{\sqrt{\text{ch}^2(x/\delta)+\varepsilon}},$$

$$\delta = \sqrt{\frac{\varkappa}{\gamma P_0^4 + \frac{\beta}{2} P_0^2}}, \quad \varepsilon = \frac{2\gamma P_0^2}{4\gamma P_0^2 + 3\beta}. \tag{1.26}$$

The corresponding distribution is shown in Fig. 2.2 which shows that at the temperatures close to the phase transition point, instead of the distribution 2, analogous to the case of the crystal with the phase transition of the second order, practically two independent distributions in the sections of alteration of polarization $-1<P/P_0<0$ and $0<P/P_0<1$ are implemented here. This behaviour of polarization in the transition layer is related to the presence of a metastable state at $P = 0$ as an intermediate state between the polar states $-P_0$ and $P_0$ (Fig. 2.3).

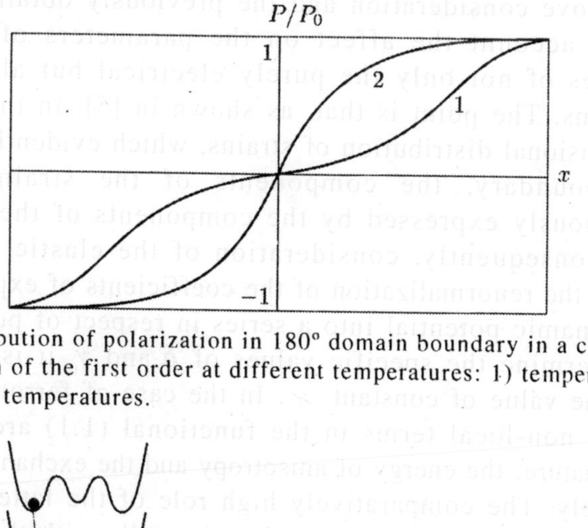


**Fig. 2.2.** Distribution of polarization in 180° domain boundary in a crystal with a phase transition of the first order at different temperatures: 1) temperatures close to $T_c$, 2) at low temperatures.

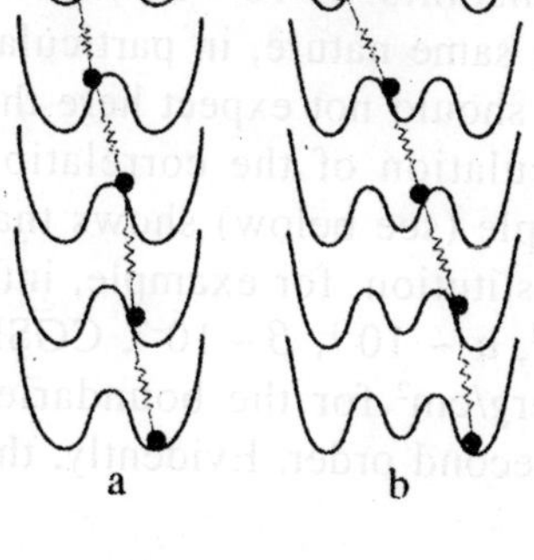


**Fig. 2.3.** Thermodynamic potential and displacement of ferroactive particles in the region of the domain boundary in crystals with the phase transition of the second (a) and first (b) order.

The surface density of the energy of the domain wall in this case taking into account the first integral (1.25) and the distribution (1.26) is

$$\begin{aligned}\gamma_0 &= 2\int_{-\infty}^{\infty}\left[\varphi(P)-\varphi(P_0)\right] = \\ &= 2\int_{-\infty}^{\infty}\left(P_0^2-P^2\right)^2\left[\frac{\gamma}{6}P^2+\frac{\beta}{4}+\frac{\gamma}{3}P_0^2\right]dx = \\ &= \frac{\varkappa P_0^2}{\delta}(1+\varepsilon)^2\int_{-\infty}^{\infty}\frac{\mathrm{ch}^2 t\,dt}{\left(\varepsilon+\mathrm{ch}^2 t\right)} = \frac{\varkappa P_0^2}{\delta}\Gamma,\end{aligned} \qquad (1.27)$$

where $\Gamma$ is some numerical factor.

As expected, equation (1.27) differs from (1.22) only by the numerical multiplier. At same time due to the finite value of $P_0$ (1.24) at $T_c$, the value of $\gamma_0$ does not convert into zero and the width of the domain wall does not diverge at $T = T_c$ for crystals with the phase transition of the first order.

The above consideration and the previously obtained solutions take into account the affect on the parameters of the domain boundaries of not only the purely electrical but also of elastic interactions. The point is that, as shown in [5], in the case of the one-dimensional distribution of strains, which evidently takes place in the boundary, the components of the strain tensor are unambiguously expressed by the components of the polarization vector. Consequently, consideration of the elastic effects leads simply to the renormalization of the coefficients of expansion of the thermodynamic potential into a series in respect of polarization.

To determine the specific values of $\delta$ and $\gamma_0$ it is necessary to specify the value of constant $\varkappa$. In the case of ferromagnetics, the local and non-local terms in the functional (1.1) are of different physical nature: the energy of anisotropy and the exchange interaction, respectively. The comparatively high role of the latter in this case results in the formation of wide domain walls with the thickness of hundreds and thousands of lattice constants ($\sim 10^3 \div 10^4$). For ferroelectrics, both of these terms are of the same nature, in particular the dipole–dipole interaction. Therefore, we should not expect here the formation of wide domain walls. The calculation of the correlation energy taking discrete structure as an example (see below) shows that $\varkappa \approx a^2$, where $a$ is the lattice constant. Substitution, for example, into (1.10) and (1.11) of the value $\varkappa \sim 10^{-15}\mathrm{cm}^2$, $a \sim 10^{-1}$, $\beta \sim 10^{-9}$, CGSE units gives the values $\delta \sim 10^{-7}\mathrm{cm}^2$, $\gamma_0 \sim 1$ erg/cm² for the boundaries in crystals with the phase transition of the second order. Evidently, the same by order of magnitude values of the width and surface energy density should also be expected for the 180° boundaries in crystals with the phase transition of the first order.

## 2.2. Structure of the 90° domain boundary in ferroelectrics in continuous approximation

In a crystal with a highly symmetric paraelectric phase, for example, in barium titanate where the resultant polarization may be oriented along any of several axes equivalent in the cubic phase, we can observe the formation of the so-called 90° domain boundaries, which separate the domains with polarization vector rotated by 90°. Let us consider the structure of these boundaries. Let us assume that in the crystallographic axes the polarization in the thickness of the adjacent domains is directed along the axes $z$ and $x$, respectively. For further considerations, it is convenient to rotate the system of coordinates by 45° around the axis $y$ so that the domain wall in the new coordinates is normal to the new axis $x$ (see Fig. 2.4).

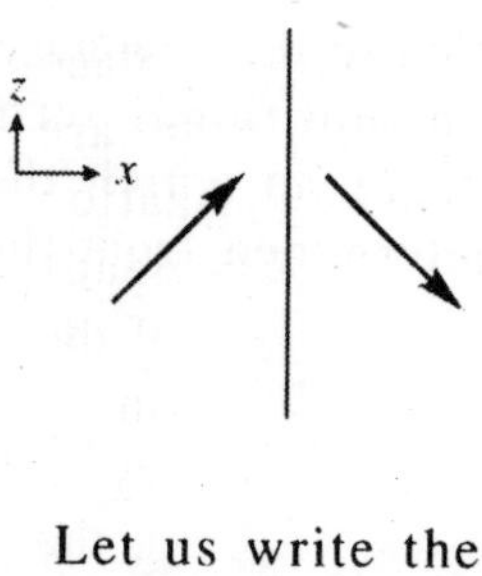


**Fig. 2.4.** 90° domain boundary in a ferroelectric crystal.

Let us write the thermodynamic potential of the ferroelectric crystal under consideration in the new coordinate system. To be more specific, let us consider the most symmetric case of barium titanate $BaTiO_3$.

It is convenient because all cases of ferroelectrics with a lower symmetry are produced from here by conversion to zero of the coefficients of the appropriate expansion terms. Assuming that the solution of the rotating type for the domain wall is energetically disadvantageous, which corresponds to the value of $P_y = 0$ for the entire boundary, the thermodynamic potential of the ferroelectric crystal including up to the sixth degree of polarization has the following form in this case:

$$\overline{\Phi}=\frac{\varkappa}{2}\left[\left(\frac{dPx}{dx}\right)^2+\left(\frac{dPz}{dx}\right)^2\right]-$$

$$-\frac{\alpha}{2}\left(P_x^2+P_z^2\right)+\frac{\beta}{4}\left(P_x^4+P_z^4\right)+ \tag{2.1}$$

$$+\frac{\beta_1}{2}P_x^2P_z^2+\frac{\gamma}{6}\left(P_x^6+P_z^6\right)+\frac{\gamma_1}{2}P_x^2P_z^2\left(P_x^2+P_z^2\right).$$

Away from the boundary, the equilibrium value of polarization in the tetragonal phase, expressed with the help of the expansion coefficients (2.1) in the rotated coordinate system, is

$$P_0=\left\{\frac{-(\beta+\beta_1)+\sqrt{(\beta+\beta_1)^2+4\alpha(\gamma+3\gamma_1)}}{(\gamma+3\gamma_1)}\right\}^{1/2}. \tag{2.2}$$

In order to write the equations of equilibrium for the polarization vector in the boundary, it is important to take into account the possibility of the presence of an internal electric field in the 90° domain boundary related to heterogeneity of distribution of the component of polarization $P_x$ in the normal direction to the boundary. This was taken into account by supplying an additional term $(\mathbf{EP})\equiv -E_xP_x$ to the thermodynamic potential (2.1), in which the field of bound charges in the boundary $E_x$ is determined from the electrostatic equation

$$\operatorname{div}\mathbf{D}=\frac{d}{dx}\left(E_x+4\pi P_x\right) \tag{2.3}$$

and has the form of

$$E_x=-4\pi\left(P_x-\frac{P_0}{\sqrt{2}}\right). \tag{2.4}$$

As a result, the minimisation of the functional $\int_{-\infty}^{\infty}\left(\overline{\Phi}-E_xP_x\right)dx$ leads to the following set of equations, which describes the distribution of the components of the polarization vector in the boundary

$$\frac{\delta\overline{\Phi}}{\delta P_z}=0,\quad \frac{\delta\overline{\Phi}}{\delta P_x}=E_x=-4\pi\left(P_x-\frac{P_0}{\sqrt{2}}\right) \tag{2.5}$$

Calculation of the variational derivatives makes it possible to write simultaneous equations (2.5) in the form:

$$\varkappa\frac{d^2P_z}{dx^2}=-\alpha P_z+\beta P_z^3+\beta_1P_x^2P_z+\gamma P_z^5+$$
$$+\gamma_1P_x^2P_z\left(P_x^2+2P_z^2\right),$$
$$\varkappa\frac{d^2P_x}{dx^2}=-\alpha P_x+\beta P_x^3+\beta_1P_xP_z^2+\gamma P_x^5+ \tag{2.6}$$
$$+\gamma_1P_xP_z^2\left(2P_x^2+P_z^2\right)+4\pi\left(P_x-\frac{P_0}{\sqrt{2}}\right).$$

The variation of the polarization component along the normal to the boundary is not large and that is why the solution of the simultaneous equations (2.6) can be found on the basis of the perturbation theory. At the same time the solution in which the component $P_x$ does not change at all, remaining everywhere equal to $P_0/\sqrt{2}$ is chosen as the zero approximation. As in the case of the 180° boundary, the component $P_z$ changes in accordance with the equation (1.12) in which the coefficients $-\alpha$, $\beta$, $\gamma$ are replaced respectively by

$$\alpha_1=-\alpha+\frac{\beta_1P_0^2}{2}+\frac{\gamma_1P_0^4}{4},\quad \beta_2=\beta+\gamma_1P_0^2,\quad \gamma\equiv\gamma, \tag{2.7}$$

and the boundary values are $P_z|_{\pm\infty}=\pm P_0/\sqrt{2}$. Consequently, in the zero approximation, the distribution of polarization in the 90° boundary is:

$$P_x=\frac{P_0}{\sqrt{2}},\quad P_z=\frac{P_0}{\sqrt{2}}\frac{\operatorname{sh}(x/\delta)}{\sqrt{\operatorname{ch}^2(x/\delta)+\varepsilon}},$$
$$\delta=2\sqrt{\frac{\varkappa}{\gamma P_0^4+\beta_2P_0^2}},\quad \varepsilon=\frac{\gamma P_0^2}{2\gamma P_0^2+3\beta_2}. \tag{2.8}$$

In this approximation, the surface density of energy $\gamma_0$, expressed by means of the quantities $\delta$ and $\varepsilon$, which are now determined by the expression (2.8), is determined here by the expression, which differs from (1.27) due to replacement of $P_0\to P_0\sqrt{2}$ formally only by the multiplier 1/2. As shown by numerical estimates in the same crystal, in particular, in $BaTiO_3$, the 90° boundary is approximately twice as wide as the 180° boundary and at the same time its surface density of the boundary energy is approximately three times less [54].

According to the calculations, the corrections in the quantities $\delta$ and $\gamma_0$ as the result of taking into account the heterogeneity of the distribution of the component $P_x$ in the boundary are actually small and can be ignored for conventional estimates. At the same time, it is very important to take into account heterogeneity of $P_x$ when investigating the problem of interaction of domain boundaries with defects where it leads to the electrostatic interaction of the charged defects with the 90° domain boundary walls. The interaction of the boundary with the defects of different nature will be discussed in the following chapter. Here, we determine the correction in the structure of the domain boundary due to the variation of the polarization component $P_x$ and find its accompanying internal electric fields in the 90° domain boundary.

The substitution of the sum $P_x + \delta P_x$, instead of $P_x$ in the second equation of (2.6), where $P_x$ is determined by the zero approximation (2.8), enables to obtain the equation for the correction $\delta P_x$ for the solution of (2.8):

$$\varkappa \frac{d^2 \delta P_x}{dx^2} = -4\pi \delta P_x = \frac{d\overline{\Phi}}{dP_x}. \tag{2.9}$$

The right hand side of (2.9) is calculated at $P_x$ and $P_z$ in the zero approximation (2.8) and is a function which differs from zero in the area of the domain boundary at distances of the order of its width $2\delta$. Since the value $\delta P_x$ also changes at the distances of $2\delta$, then according to the order of magnitude the correlation term is equal to

$$\varkappa \frac{d^2 \delta P_x}{dx^2} \sim \varkappa \frac{\delta P_x}{\delta^2} \sim \left(\gamma P_0^4 + \beta_2 P_0^2\right) \delta P_x \ll -4\pi \delta P_x \tag{2.10}$$

Therefore, in equation (2.9) the first term can be rejected and written in a simpler form

$$\delta P_x = -\frac{1}{4\pi} \frac{\partial \overline{\Phi}}{\partial P_x} \tag{2.11}$$

Substitution of the right-hand side in (2.11) taking into account the equations (2.1) and (2.8), and also the substitution of the ratio between the coefficients of expansion and the equilibrium value of $P_0$ (2.2), enables to obtain the correction $\delta P_x$ in the explicit form:

$$\delta P_x = \frac{1}{4\pi} \left\{ \frac{\left(\beta_1 + \gamma_1 P_0^2\right) P_0^3}{2\sqrt{2}} \left[ 1 - \frac{\mathrm{sh}^2 (x/\delta)}{\left(\mathrm{ch}^2 (x/\delta) + \epsilon\right)} \right] + \frac{\gamma_1 P_0^5}{4\sqrt{2}} \left[ 1 - \frac{\mathrm{sh}^4 (x/\delta)}{\left(\mathrm{ch}^2 (x/\delta) + \epsilon\right)^2} \right] \right\}. \tag{2.12}$$

Equation (2.12) shows that correction $\delta P_x$ in fact differs from zero in the area with the size of the order of $2\delta$. It is symmetric and reaches the maximum value (Fig. 2.5) equal to

$$\delta P_x\big|_{\max} = \frac{\left(\beta_1 + \frac{3}{2}\gamma_1 P_0^2\right) P_0^3}{8\sqrt{2\pi}} \tag{2.13}$$

in the centre of the boundary at $x = 0$.

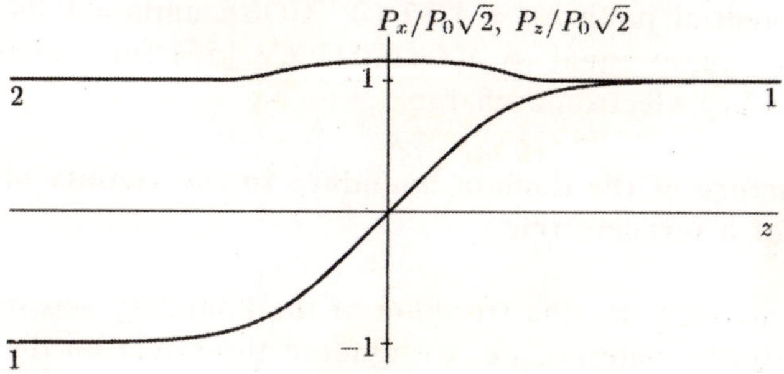


**Fig.2.5.** The variation of the polarization vector in the 90° domain boundary: 1,2 — polarization components $P_z$ and $P_x$ in relation to the position in the boundary.

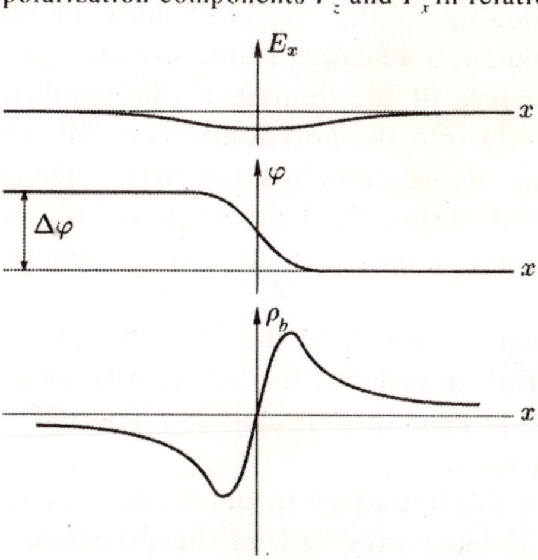


**Fig.2.6.** The distribution of the field, potential and bound charges in the area of the 90° domain boundary.

The strength of the internal electric field in the boundary, according to (2.4), is

$$E_x = -4\pi\delta P_x. \tag{2.14}$$

In accordance with (2.14) and (2.12), the distribution of the potential in the boundary is asymmetric and typical of the double electrical layer formed by the separation of the bound electric charges distributed with the density $\rho_b = -\text{div}\ \mathbf{P} = d\delta P_x/dx$ (see Fig. 2.6).

As indicated by Fig. 2.4, in the area of the boundary the

electrostatic potential shows a 'jump', the value of which is approximately equal to

$$\Delta\varphi \simeq E_{max} 2\delta = \frac{\left(\beta_1 + \frac{3}{2}\gamma_1 P_0^2\right) P_0^3}{\sqrt{2}} \delta. \tag{2.15}$$

The numerical estimates obtained from equation (2.15) for $\delta = 2 \cdot 10^{-7}$ cm, $\beta_1 = 10^{-12}$, $\gamma_1 = 10^{-22}$, $P_0 = 7.8 \cdot 10^4$ CGSE units leads to the potential jump $\Delta\varphi = 1.37 \cdot 10^{-4}$ CGSE units = 0.041 V or to the energy jump equal to $\Delta\mathcal{U} = 0.041$ eV [55] for a charge equal to elementary electronic charge.

## 2.3. Structure of the domain boundary in the vicinity of the surface of a ferroelectric

In previous sections, the structure of the boundary was determined for an infinite material, i.e. we ignored the effect on the structure of the boundary produced by the depolarizing fields, formed in the vicinity of the surface of the crystal of finite dimensions. The results obtained here are applicable to the so-called bulk of the material, i.e. to sections of the domain boundaries whose distance from the surface is greater than the width of the domain $d$ (the depth of penetration of the depolarizing field into the polydomain crystal, see chapter 1), and also for sections of the domain boundaries located closer to the surface if the depolarizing field is compensated, for example, as a result of volume conductivity or charges on the surface states.

In the absence of this compensation due to the influence of the depolarizing field the polarization is reduces in the vicinity of the surface of the ferroelectric and, in principle, it can also have effect on the structure of the domain wall.

To determine the structure of the boundary in the vicinity of the surface of the ferroelectric, it is necessary to find the distribution of polarization in this part of the crystal. Even for the laminated domain structure due to the two-dimensional nature of the problem being solved and the nonlinear relation between polarization and the depolarizing field, the problem under consideration is not solved in the general form away from the Curie point. In order to simplify it, the initial thermodynamic potential for a ferroelectric with the phase transition of the second order $\overline{\Phi} = -\frac{\alpha}{2}P^2 + \frac{\beta}{4}P^4$ is replaced by a set of two shifted parabolas $\overline{\Phi}=\overline{\Phi}_0=\frac{\alpha(P\pm P_0)^2}{2}$, $P_0^2=\alpha/\beta$, obtained as a result of the expansion of the initial potential $\overline{\Phi}$ into a series in the vicinities of its minima.

Consequently, the set of the equations describing the distribution of polarization and the electric field in the ferroelectric crystal with the polar axis $z$, is written in the form:

$$2\alpha\left(P-P_0(x)\right)-\varkappa\frac{d^2P}{dx^2}=\frac{\partial\varphi}{\partial z},$$
$$\frac{\partial^2\varphi}{\partial x^2}+\frac{\partial^2\varphi}{\partial z^2}=4\pi\frac{dP}{dz}. \tag{3.1}$$

In (3.1), as well as in section 1.1, $P_0(x)$ is the odd periodic function, determined in the period $(-d,d)$ as

$$P_0\operatorname{sign}x=P_0\begin{cases}1, & 0<x<d,\\ -1, & -d<x<0.\end{cases} \tag{3.2}$$

(the 180° domain structure is discussed) and to simplify considerations it is assumed that $\varepsilon_x = 1$.

Let us find the solution of the system (3.1)–(3.2) in the form of expansion into a Fourier series in respect of axes $x$ where the expansion coefficients depend on the coordinate along the polar axis $z$. Taking into account the symmetry of the problem, we obtain

$$P(x,z)=\sum_{n=0}^{\infty}P_{2n+1}(z)\sin\frac{(2n+1)\pi x}{d},$$
$$\varphi(x,z)=\sum_{n=0}^{\infty}\varphi_{2n+1}(z)\sin\frac{(2n+1)}{d}\pi x. \tag{3.3}$$

From the first equation of system (3.1), the relation between the Fourier coefficients of expansion of polarization and the potential has the form of

$$P_{2n+1}(z)=\frac{8P_0\alpha}{\pi(2n+1)\left[2\alpha+\varkappa(2n+1)^2\pi^2/d^2\right]}-$$
$$-\frac{1}{\left[2\alpha+\varkappa(2n+1)^2\pi^2/d^2\right]}\frac{\partial\varphi_{2n+1}}{\partial z}. \tag{3.4}$$

On the basis of Laplace's equation, the distribution of the potential in the crystal and outside it is as follows:

$$\varphi_{2n+1}(z \geq 0) = A\exp\left[-\frac{k_{2n+1}z}{\sqrt{\varepsilon_z(k)}}\right],$$
$$\varphi_{2n+1}(z \leq 0) = B\exp\left[k_{2n+1}z\right],$$
$$k_{2n+1} = (2n+1)\pi/d, \tag{3.5}$$
$$\varepsilon_z(k) = 1 + \frac{4\pi}{2\alpha + \varkappa(2n+1)^2\pi^2/d^2}.$$

The solution of (3.5) should satisfy the boundary conditions

$$\varphi+(0) = \varphi-(0), \quad \frac{\partial\varphi+}{\partial z} - \frac{\partial\varphi-}{\partial z} = 4\pi P(0), \tag{3.6}$$

which enables us to determine the coefficients explicitly

$$A = B, \quad A = -\frac{4\pi P_{2n+1}(0)}{k_{2n+1}\left(1 + 1/\sqrt{\varepsilon_z(k)}\right)}. \tag{3.7}$$

Now, on the basis of (3.4) and (3.7) the final expression for the Fourier coefficients of polarization expansion on the surface of the ferroelectric is:

$$P_{2n+1}(0) = \frac{8P_0\alpha\left(\sqrt{\varepsilon_z(k)} + 1\right)}{\pi^2(2n+1)\left[\left(2\alpha + \varkappa k_{2n+1}^2\right)\left(\sqrt{\varepsilon_z(k)} + 1\right) + 4\pi\right]}. \tag{3.8}$$

or approximately

$$P_{2n+1}(0) \simeq \frac{8P_0\alpha\sqrt{\varepsilon_z(k)}}{4\pi^2(2n+1)} = \frac{4\sqrt{\pi}\alpha P_0}{\pi^2(2n+1)\sqrt{2\alpha + \varkappa k_{2n+1}^2}}. \tag{3.9}$$

Consequently

$$P(x,0) = \sum_{n=0}^{\infty} \frac{4P_0\alpha\sqrt{\pi}}{\pi^2(2n+1)\sqrt{2\alpha + \varkappa k_{2n+1}^2}} \sin\frac{\pi(2n+1)x}{d}. \tag{3.10}$$

To determine the width $\delta$ of the domain boundary in the vicinity of the surface of the ferroelectric crystal let us determine the value $P(x,0)$ in the middle of the domain and the value of the derivative $\partial P/\partial x$ in the centre of the domain wall. Polarization on the surface of the crystal in the middle of the distance between the domain boundaries is:

$$P(d/2,0) \simeq \frac{4P_0}{\pi}\sqrt{\frac{\alpha}{2\pi}}. \tag{3.11}$$

Similarly, the derivative

$$\frac{\partial P(0,0)}{\partial x} = \frac{4\sqrt{\alpha}}{\sqrt{2\pi}}\frac{P_0}{d}\sum_{n=0}^{\infty}\frac{1}{\sqrt{1+\frac{\varkappa}{2d}\frac{\pi^2(2n+1)^2}{d^2}}} \simeq \frac{2\alpha P_0}{\pi\sqrt{\pi}\sqrt{\varkappa}}. \tag{3.12}$$

Then the width of the domain wall in the vicinity of the surface of the ferroelectric is

$$\delta = \frac{P(d/2,0)}{\partial P(0,0)/\partial x} = \sqrt{\frac{2\varkappa}{\alpha}}. \tag{3.13}$$

As indicated by (3.13), the width of the domain wall in the vicinity of the surface of the ferroelectric coincides accurately with its width in the bulk of the material (1.21) [56]. It should be mentioned that the obtained result could have been predicted to a certain degree because the previously derived equation for the width of the wall in the bulk (1.21) does not depend explicitly on polarization $P$, and consequently, should not change with the alteration of polarization that takes place in the vicinity of the surface of the crystal.

## 2.4. Structure of the interphase boundaries in ferroelectrics

If in the case of domain walls in ferroelectrics it is possible to discuss their preferential orientation determined mostly by symmetry considerations, then for the interphase boundaries (because of the obviously closed surface restricting the nucleus of the new phase), it is necessary to consider boundaries with an arbitrary orientation. Below, we consider the boundaries of two qualitatively different types: parallel and normal to the direction of spontaneous polarization. The boundary with the arbitrary orientation may be constructed with the using of the mentioned boundaries as a basis.

Let us consider a flat interphase boundary with the normal to it coinciding with axis $x$, separating the non-polar state and the polar state with the polarization oriented parallel to the plane of the boundary, for example, along axis $z$.

Let us use the conventional expansion of the thermodynamic potential

$$\overline{\Phi} = \frac{\varkappa}{2}\left(\frac{dP}{dx}\right)^2 + \frac{\alpha}{2}P^2 + \frac{\beta}{4}P^4 + \frac{\gamma}{6}P^6, \tag{4.1}$$

the coefficients of which in the problem under consideration satisfy the certain conditions. On the one hand, since we consider the stable existence of phases (polar and non-polar), the values of the thermodynamic potential for these states should coincide. This is expressed in the following condition

$$\alpha(T_c)+\frac{\beta}{2}P_0^2+\frac{\gamma}{3}P_0^4=0 \tag{4.2}$$

On the other hand, from the equation of equilibrium in the ferroelectric area it follows that

$$\alpha(T_c)+\beta P_0^2+\gamma P_0^4=0 \tag{4.3}$$

Evidently, the system of equations (4.2), (4.3) give a specific ratio between the coefficients of expansion which, in particular, can be written in the following form:

$$\frac{\gamma}{3}=\frac{\beta^2}{16\alpha(T_c)}, \quad P_0^2=\frac{3|\beta|}{4\gamma} \tag{4.4}$$

The equation resulting from (4.1) describing the distribution of polarization in the interphase boundary has the conventional form

$$\varkappa\frac{d^2P}{dx^2}=\alpha(T_c)+\beta P^3+\gamma P^5 \tag{4.5}$$

with the following boundary conditions in this case $P(-\infty) = 0$, $P(+\infty) = P_0$. The first integral of this equation, presented in section 2.1, in our case is simplified taking into account (4.4) due to reduction of the last two terms in the square brackets of the right hand part. Taking into account the ratio $\gamma/6 = \beta^2/32\alpha(T_c) = \alpha(T_c)/2\,P_0^4$ it is written in the form

$$\delta^2\left(\frac{dP}{dx}\right)^2=P^2\left(P^2-P_0^2\right)^2/P_0^4, \quad \delta=\sqrt{\varkappa/\alpha(T_c)}. \tag{4.6}$$

The separation of the variables and the subsequent integration in (4.6) gives the following distribution of polarization in the area of the interphase boundary [57]:

$$P(x)=\frac{P_0}{\sqrt{\exp[-2x/\delta]+1}}. \tag{4.7}$$

The surface density of the energy of the interphase boundary of the given orientation, taking into account the first integral, is:

$$\gamma_0 = \varkappa \int_{-\infty}^{\infty} \left( \frac{dP}{dx} \right)^2 dx = \frac{\varkappa}{4\delta} P_0^2, \tag{4.8}$$

which is considerably lower than the energy of the 180° domain boundary (see equation (1.22)), with the same values of the quantities included in the equation.

The interphase boundary of the considered orientation is evidently not charged. In contrast to it in the boundary with the normal to its surface coinciding with the polar axes the alteration of polarization is associated with the formation of bound electric charges. Therefore, when investigating the structure of such a boundary it is essential to take into account the interaction of depolarizing fields of these charges with polarization and also the possibility of their screening by carriers of some kind [58]. From the viewpoint of calculations this leads to the situation, in which instead of the variational derivative being equal to zero $\delta\overline{\Phi}/\delta P = 0$ and leading to equation (4.5), the mentioned derivative is equated here to the strength of the depolarizing field of the bound charges at the boundary $\delta\overline{\Phi}/\delta P = E$. In this case, field $E$ is discovered from conventional electrostatic equations.

Assuming that the screening of polarization in the area of the interphase boundary is carried out by a non-degenerate gas of the electrons and holes with the carrier concentration $n_0$, the mentioned set of equations, in which we assume $\varepsilon_x = 1$ to simplify calculations and taking into account (4.4), can be written in the following form [59]:

$$-\frac{d\varphi}{dz} = \frac{\delta\overline{\Phi}}{\delta P}, \quad \overline{\Phi} = \frac{\varkappa}{2}\left(\frac{dP}{dz}\right)^2 + \frac{\alpha(T_c)}{2P_0^4}\left(P_0^2 - P^2\right)P^2,$$
$$-\frac{d^2\varphi}{dz^2} + 4\pi\frac{dP}{dz} = 4\pi\left(-2n_0 e \operatorname{sh}\frac{e\varphi}{kT}\right) \simeq -\frac{8\pi n_0 e^2}{kT}\varphi = -\frac{\varphi}{\lambda^2}. \tag{4.9}$$

As indicated by (4.9), we obtain different results depending on the degree of screening. In the absence of screening, i.e. at $n_0 \to 0$, $\lambda \to \infty$, the value of the depolarizing field according to (4.9) is equal to $E = -4\pi P$, i.e. to the value considerably higher than the thermodynamic coercive field. In this field, the polarization that creates the field is unstable, i.e. it should spontaneously reverse. Therefore, there is no sense in discussion of the structure of the flat interphase boundary perpendicular to the polarization vector in the absence of such screening.

Let us assume that a sufficient degree of screening is present and the screening length $\lambda$ is sufficiently small so that the inequality $d^2\varphi/dz^2 << \varphi/\lambda^2$ is fulfilled between the terms in the second equation of (4.9). Consequently, the value of the depolarizing field is

$$-\frac{d\varphi}{dz} = 4\pi\lambda^2 \frac{d^2P}{dz^2} \tag{4.10}$$

and the equation describing the distribution of polarization in the area of the interphase boundary has the form:

$$\left(\varkappa + 4\pi\lambda^2\right)\frac{d^2P}{dz^2} = \frac{d}{dP}\left(\frac{\alpha(T_c)}{2P_0^4}\left(P_0^2 - P^2\right)P^2\right), \tag{4.11}$$

This equation differs from equation (4.5) only by the new constant $\tilde{\varkappa} = \varkappa + 4\pi\lambda^2$ of the correlation term. Therefore, we can immediately write the distribution of polarization in the area of the interphase boundary of the given orientation:

$$P(x) = \frac{P_0}{\sqrt{\exp\left[-2x/\sqrt{\delta_C^2 + \delta_D^2}\right] + 1}},$$
$$\delta_C = \sqrt{\frac{\varkappa}{\alpha(T_c)}}, \quad \delta_D = \sqrt{\frac{4\pi\lambda^2}{\alpha(T_c)}}. \tag{4.12}$$

The surface density of the energy of the interphase boundary with given orientation is formed by polarization, its interaction with the depolarizing field and the electronic subsystem. The functional, the variation of which gives the system (4.9) has the form

$$\gamma_0 = \int\left[\bar{\Phi} - \frac{1}{8\pi}\left(\frac{\partial\varphi}{\partial z}\right)^2 - 4Tn_0\mathrm{sh}^2\frac{e\varphi}{kT} + P\frac{d\varphi}{dz}\right]dz. \tag{4.13}$$

The value $\gamma_0$, determined as the result of substituting the distribution (4.12) into (4.13) just represents the energy of the interphase boundary. Taking into account the first integration of the system (4.9) the integral (4.13) can be rewritten in the form

$$\gamma_0 = \int\left[\varkappa\left(\frac{dP}{dz}\right)^2 - \frac{1}{4\pi}\left(\frac{d\varphi}{dz}\right)^2 + P\frac{d\varphi}{dz}\right]dz. \tag{4.14}$$

If $\delta_C << \delta_D$, the main contribution to (4.14) comes from the correlation term and then the interphase boundary energy is determined by the already found expression (4.8). However, if $\delta_C >> \delta_D$ then on the contrary, the first term in (4.14) can be ignored. In the approximation $d^2\varphi/dz^2 \ll \varphi/\lambda^2$ in the case under consideration, the field $-d\varphi/dz \gg 4\pi P$ and, consequently, the main contribution to the integral (4.14) comes from the last term. Substituting into this integral distribution (4.12) at $\delta_C \gg \delta_D$ and taking (4.10) into account, we obtain

$$\gamma_0 = \int_{-\infty}^{\infty} P\frac{d\varphi}{dz}dz = -4\pi\lambda^2 \int_{-\infty}^{\infty} P\frac{d^2P}{dz^2} =$$
$$= 4\pi\lambda^2 \int_{-\infty}^{\infty}\left(\frac{dP}{dz}\right)^2 dz = \frac{\pi\lambda^2}{\delta_D}P_0^2 = \frac{\lambda P_0^2}{2}\sqrt{\pi\alpha(T_c)}. \quad (4.15)$$

As expected, expression (4.15) is almost identical with similar expressions (1.22) and (4.8) with the accuracy to the change of the meaning of the thickness of the transition layer from the correlation length to the Debye screening length. Since in most cases, $\delta_C \gg \delta_D$, the energy of the charged interphase boundary is also higher than that of the uncharged interphase boundary, even if screening is taken into account.

## 2.5. Structure of the domain boundaries in improper ferroelectrics and ferroelectrics with an incommensurate phase

In improper ferroelectrics, the polarization that occurs at phase transition is not an order parameter, i.e. it does not describe the alteration of symmetry that takes place during the phase transition. In this case, the ratio of the polarization with the order parameter is non-linear and, due to this the domains do not coincide with each other in respect of the order parameter and the polarization vector. Therefore, the physical nature of the domain walls also differs here depending on the nature of the domains, which it separates. Some domain walls, which separate the domains with different polarization vectors, are ferroelectric domain walls. Others, in which polarization in the separated domains is the same, are the so-called antiphase domain boundaries or simply antiphase boundaries [60].

This will be shown by the example of an improper ferroelectric crystal with the symmetry of gadolinium molybdate $Gd_2(MnO_4)_3$. The thermodynamic potential of this crystal in the presence of the one-dimensional heterogeneity in this case is as follows

$$\Phi = \int \left\{ \frac{1}{2} \varkappa \left[ \left( \frac{dq_1}{dx} \right)^2 + \left( \frac{dq_2}{dx} \right)^2 \right] - \frac{1}{2} \alpha \left( q_1^2 + q_2^2 \right) + \right.$$
$$\left. + \frac{1}{4} \beta \left( q_1^4 + q_2^4 \right) + \frac{1}{2} \gamma' q_1^2 q_2^2 + \xi q_1 q_2 P + \frac{1}{2\chi_0} P^2 \right\} dx. \tag{5.1}$$

The minimization of potential $\Phi$ in respect of $P$ gives the non-linear ratio

$$P = -\xi \chi_0 q_1 q_2, \tag{5.2}$$

whose substitution into (5.1) results in the renormalization of coefficient $\gamma$ in the expansion of potential $\Phi$ into a series in respect of the two-component order parameter $(q_1, q_2)$. Consequently, potential $\Phi$ (5.1) is rewritten in the following form

$$\Phi = \int \left\{ \frac{\varkappa}{2} \left[ \left( \frac{dq_1}{dx} \right)^2 + \left( \frac{dq_2}{dx} \right)^2 \right] - \frac{1}{2} \alpha \left( q_1^2 + q_2^2 \right) + \frac{\beta}{4} \left( q_1^4 + q_2^4 \right) + \frac{\gamma}{2} q_1^2 q_2^2 \right\} dx. \tag{5.3}$$

According to (5.3), the distribution of the order parameter is described here by the set of equations

$$\varkappa \frac{d^2 q_1}{dx^2} = -\alpha q_1 + \beta q_1^3 + \gamma q_1 q_2^2,$$
$$\varkappa \frac{d^2 q_2}{dx^2} = -\alpha q_2 + \beta q_2^3 + \gamma q_1^2 q_2. \tag{5.4}$$

In a homogeneous state at $-\beta < \gamma < \beta$, the following states are stable (Fig. 2.7) [61]:

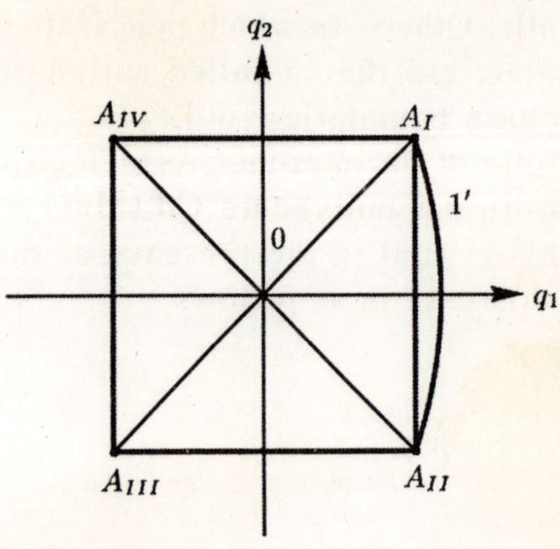


**Fig.2.7.** Distribution of the stable states of the thermodynamic potential (5.3) at $-\beta < \gamma < \beta$ (points $A_I$, $A_{II}$, $A_{III}$, $A_{IV}$) and possible domain boundaries in the system under consideration (sides and diagonals of the square, other curves).

$$\begin{array}{lll} I, & q_1 = q_2 = q_0 = \sqrt{\alpha/(\beta+\gamma)}, & P = -x_0 \xi q_0^2 = -P_0, \\ II, & -q_1 = q_2 = q_0, & P = P_0, \\ III, & -q_1 = -q_2 = q_0, & P = -P_0, \\ IV, & q_1 = -q_2 = q_0, & P = -P_0. \end{array} \quad (5.5)$$

At $\gamma > \beta$ the pattern of the stable states is shown in Fig. 2.8.

$$\begin{array}{ll} I, & q_1 = q_0 = \sqrt{\alpha/\beta},\ q_2 = 0, \\ II, & q_1 = 0,\ q_2 = q_0, \\ III, & q_1 = -q_0,\ q_2 = 0, \\ IV, & q_1 = 0, q_2 = -q_0,\quad P_I = P_{II} = P_{III} = P_{IV} = 0. \end{array} \quad (5.6)$$

The stability of the states (5.5) or (5.6) is determined by comparing the values of the thermodynamic potential in points $A\left(\Phi = -\dfrac{\alpha^2}{2(\beta+\gamma)}\right)$ and points B $\left(\Phi = -\dfrac{\alpha^2}{4\beta}\right)$, respectively.

The transition from one stable state (domain) to another within the limits of each of the diagrams (5.5) or (5.6) represents domain walls in the material under consideration. These walls correspond to the lines (the sides, the diagonals of the square or other curves) on the graphs. As indicated by the distribution (5.5), all consecutive transitions $A_I \Leftrightarrow A_{II}$, $A_{II} \Leftrightarrow A_{III}$, $A_{III} \Leftrightarrow A_{IV}$, $A_{IV} \Leftrightarrow A_I$ represent ferroelectric domain walls whereas the transitions $A_I \Leftrightarrow A_{III}$, $A_I \Leftrightarrow A_{IV}$ are antiphase boundaries. For the distribution (5.6), none of the stable states is linked with the formation of polarization and, consequently, all the transitions between them (both the sides and diagonals of the square, and the other curves in Fig. 2.8) represent only antiphase boundaries.

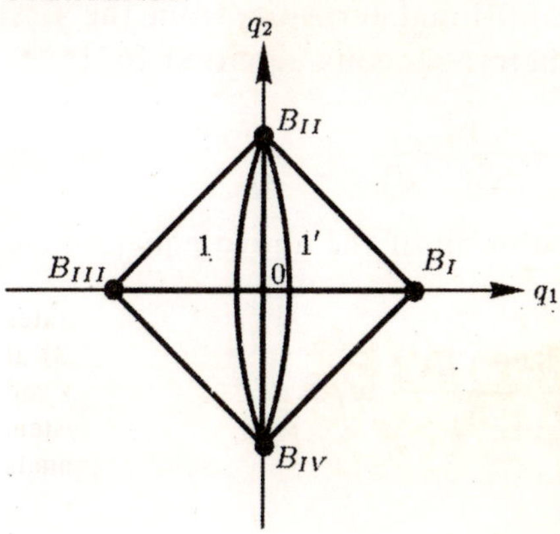


**Fig.2.8.** Distribution of stable states at $\gamma > \beta$ (points $B_I$, $B_{II}$, $B_{III}$, $B_{IV}$) and possible domain boundaries (sides and diagonals of the square, other curves).

The process of transition itself from one stable state to another in the antiphase boundaries can be linked either with the onset of polarization (the sides of the square and curves *1* and *1'* in Fig.2.8), or with its variation – the diagonals of the square in Fig.2.7), or it takes place without appearance of polarization at all (the diagonals of the square in Fig.2.8).

The specific distribution of the order parameter (its components $q_1$, $q_2$) together with the possible change of the polarization in the boundaries described above is determined by set of equations (5.4), ratio (5.2) and the corresponding boundary conditions. Unfortunately, in the general case the analytical solution of system (5.4) has not been found and solutions exist only for the individual particular cases. For example, at $q_1(x) = q_2(x)$ the solution corresponding to the antiphase boundary $A_I O A_{III}$ has the form:

$$q_1 = q_2 = q_0 \text{th}(x/\delta), \quad q_0 = \sqrt{\frac{\alpha}{\beta+\gamma}}, \quad \delta = \sqrt{\frac{2\varkappa}{\alpha}}. \tag{5.7}$$

At $\gamma = 0$ transition from $A_I$ to $A_{II}$ takes place along the side of the square, i.e. in accordance with the distribution

$$q_2(x) = q_0 \text{th}(x/\delta), \quad q_1(x) = q_0, \tag{5.8}$$

and the transition $B_{II} \Leftrightarrow B_{IV}$ by means of the single-component wall, i.e. along the straight line in the scheme in Fig.2.8

$$q_2 = q_0 \text{th}(x/\delta), \quad q_1 = 0. \tag{5.9}$$

At $\gamma \neq 0$, the solution of system (5.4) can be found by numerical calculations, the variational method, or (in the presence of a small parameter) using the perturbation theory.

Let us find using the last method the solution of set (5.4) for the transition $A_I \Leftrightarrow A_{II}$ in particular. Let us consider the case $\gamma/\beta << 1$. We are going to find in this case the solution in the form of $q_2 = q_0 \text{th}(x/\delta)$, $q_1(x) = q_0 + \delta q_1$. For additional term $\delta q_1$ from the first of equations in (5.4), we have a heterogeneous equation [62]***

$$\varkappa \frac{d^2 \delta q_1}{dx^2} - 2\alpha \delta q_1 = -\frac{\gamma q_0^3}{\text{ch}^2(x/\delta)}. \tag{5.10}$$

Using the Green function, the solution of the above equation can be written as follows

$$\delta q_1(x) = -\frac{\gamma q_0}{2\beta} \int_{-\infty}^{\infty} \frac{\exp(-2|x-x'|/\delta)}{\text{ch}^2(x'/\delta)} dx'/\delta. \tag{5.11}$$

It is evident that solution $\delta q_1(x)$ is symmetric, converts to zero at $x\to\pm\infty$, and at $x = 0$ has a maximum whose value from (5.11) is equal to

$$\delta q_1(x=0) \simeq 0.19\frac{\gamma q_0}{\beta}. \qquad (5.12).$$

.Thus, the transition in the boundary $A_I \Leftrightarrow A_{II}$ takes place along the curve *1'*, Fig.2.7, witn the value of the deviation from the straight line equal to (5.12).

The energy value of the appropriate boundary is obtained by substituting the found solutions into (5.1) and in this case is approximately equal to:

$$\gamma_0 = \frac{\alpha^{3/2}\varkappa^{1/2}}{\beta} \simeq \alpha q_0^2 \delta, \qquad (5.13)$$

i.e. it is described by the expression similar to the equation for $\gamma_0$ for the domain boundary in conventional ferroelectrics.

Investigation of the structure of domain walls in ferroelectrics with an incommensurate phase is of specific interest. It has been established that the incommensurate phase characterized by the modulated distribution of the order parameter occurs in the systems with competing interactions whose presence fosters the heterogenous distribution of the order parameter. It is expected that this feature of these crystals should be reflected in the domain structure and in the commensurate ferroelectric phase.

One of the simplest forms of the volume density of the thermodynamic potential in such crystals has the form

$$\overline{\Phi}(x) = \frac{\alpha}{2}P^2 + \frac{\beta}{4}P^4 + \frac{\varkappa}{2}\left(\frac{dP}{dx}\right)^2 + \frac{\sigma}{2}\left(\frac{d^2P}{dx^2}\right)^2 + \frac{\eta}{2}P^2\left(\frac{dP}{dx}\right)^2. \qquad (5.14)$$

Assuming for simplicity that $\eta = 0$, the equation for the distribution of the polarization vector from the variation of the functional $\int\overline{\Phi}(x)dx$ is written in the form

$$\sigma\frac{d^4P}{dx^4} - \varkappa\frac{d^2P}{dx^2} + \alpha P + \beta P^3 = 0. \qquad (5.15)$$

The numerical solution of this equation with the boundary conditions $P = \pm P_0$ at $x \to \pm\infty$ in the area of stability of the commensurate ferroelectric phase leads to the distribution $P(x)$ in the boundary shown in Fig.2.9.

The characteristic distinguishing feature of this distribution is the approach of polarization to the equilibrium value after oscillation as

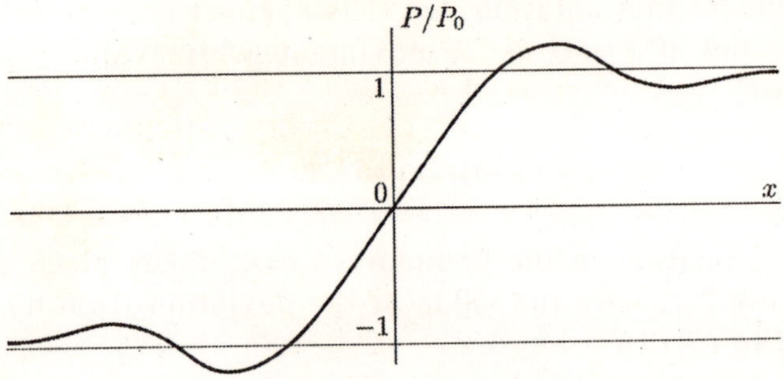


**Fig.2.9.** Distribution of polarization in a domain wall in a commensurate phase of a ferroelectric crystal, preceded by the incommensurate phase.

the function of coordinate $x$. The presence of sections in which polarization $P$ exceeds the equilibrium value $P_0$ and the oscillating nature of approach of polarization to equilibrium evidently reflects the tendency for spatial modulation in the investigated materials [63].

## 2.6. Phase transitions in domain walls in ferroelectrics and related materials

In the case of a multicomponent order parameter or several order parameters, their distribution in the area of the domain wall allows several options. With the alteration of the conditions in which the investigated material is kept in particular its temperature or pressure qualitative changes may take place in the structure of the domain wall associated with the transition from one option of the structural distribution of the order parameter in the wall to another, whereas the structure and symmetry of the material in the bulk of the domain remain unchanged. Such changes take place in the form of a phase transition and are referred to as phase transitions in the domain walls.

The phase transition in the domain wall, induced by temperature changes, was observed for the first time in $D_yFeO_3$ [64]. The transition in the domain wall, induced by the external magnetic field, was observed in $CuCl_2{\times}2H_2O$ [65] and $(C_2H_5NH_3)_2CuCl_2$ [66].

The special attention to the phase transitions in the domain walls that has been paid in recent years is undoubtedly associated with the discovery of high-temperature conductivity and with the fact [67, 68] that the phase transition temperature in the superconducting state in twin boundaries is higher than the temperature of such transitions in the bulk of the material. In other words, the domain or twin boundaries are treated as a natural factor contributing to the increase the transition temperature to the superconducting state.

Let us consider the phase transition in the domain wall by the example of the material with two single-component ordering parameters $\eta$ and $\varphi$ in the first place. The simplest thermodynamic potential, describing such a system, has the form of

$$\Phi = \int\left\{\frac{\varkappa_1}{2}\left(\frac{d\eta}{dx}\right)^2 + \frac{\varkappa_2}{2}\left(\frac{d\varphi}{dx}\right)^2 - \frac{\alpha_1}{2}\eta^2 + \right.$$
$$\left. + \frac{\beta_1}{4}\eta^4 - \frac{\alpha_2}{2}\varphi^2 + \frac{\beta_2}{4}\varphi^4 + \frac{\gamma}{2}\varphi^2\eta^2\right\} dx. \tag{6.1}$$

Depending on the ratio between the coefficients, this potential permits four homogeneous phases (Fig. 2.10).

In phase I, $\eta = \varphi = 0$. In phase II, $\eta \neq 0$, $\varphi = 0$. This phase exists in the area from $-\alpha_1 < 0$ to $-\alpha_1 < \alpha_2\beta_1/\gamma$. In phase III $\eta = 0$, $\varphi \neq 0$. This phase exists in the area from $-\alpha_2 < 0$ to $-\alpha_1 > -\alpha_2\gamma/\beta_2$. And, finally phase IV, where $\eta \neq 0$, $\varphi \neq 0$ exists between the lines 1 and 2, i.e. from $-\alpha_1 < -\alpha_2\gamma/\beta_2$ to $-\alpha_1 < -\alpha_2\beta_1/\gamma$.

The distribution of the order parameters in sections with heterogeneous $\eta$ and $\varphi$, in particular, in the region of the domain boundaries is described by the set of equations which, as usual, is obtained by varying the potential (6.1) in respect of $\eta$ and $\varphi$:

$$\varkappa_1\frac{d^2\eta}{dx^2} = -\alpha_1\eta + \beta_1\eta^3 + \gamma\eta\varphi^2,$$
$$\varkappa_2\frac{d^2\varphi}{dx^2} = -\alpha_2\varphi + \beta_2\varphi^3 + \gamma\eta^2\varphi. \tag{6.2}$$

For phase II, the set of equation (6.2) has a conventional solution describing the single-parameter domain wall:

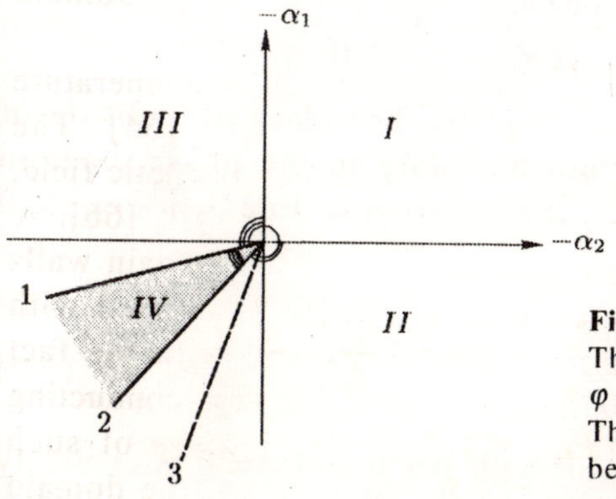


**Fig.2.10.** Phase diagram for potential (6.1). The single-parameter wall with $\eta \neq 0$, $\varphi = 0$ forms to the right of the line 3. The domain wall with $\eta \neq 0$, $\varphi \neq 0$ forms between the lines 2 and 3.

$$\eta(x)=\eta_0 \text{th}\frac{x}{\delta}, \qquad \eta_0=\sqrt{\frac{\alpha_1}{\beta_1}},$$
$$\delta=\sqrt{\frac{2\varkappa_1}{\alpha_1}}, \qquad \varphi(x)=0. \tag{6.3}$$

As in the case of (1.9), the problem of determination of the stability of this solution is reduced to investigation of the spectrum of the eigenvalues of the set of equations

$$-\varkappa_1\frac{d^2\psi_n'}{dx^2}+\left(-\alpha_1+3\beta_1\eta^2+\gamma\varphi^2\right)\psi_n'=\varepsilon_n\psi_n',$$
$$-\varkappa_2\frac{d^2\psi_n}{dx^2}+\left(-\alpha_2+3\beta_2\varphi^2+\gamma\eta^2\right)\psi_n=\lambda_n\psi_n. \tag{6.4}$$

The substitution of the solution under investigation (6.3) into (6.4) makes it possible to present each of the equations of set (6.4) in the form of a Schrödinger equation with the potential of $V(x)\sim \text{ch}^{-2}(x/\delta)$:

$$-\varkappa_1\frac{d^2\psi_n'}{dx^2}+\alpha_1\left[2-\frac{3}{\text{ch}^2(x/\delta)}\right]\psi_n'=\varepsilon_n\psi_n',$$
$$-\varkappa_2\frac{d^2\psi_n}{dx^2}+\left(-\alpha_2+\gamma\eta_0^2\right)\psi_n-\frac{\gamma\eta_0^2}{\text{ch}^2(x/\delta)}=\lambda_n\psi_n. \tag{6.5}$$

The condition of stability of the solution of (6.3) is the non-negativeness of all eigenvalues of set (6.5). The use of the already known solution of equations of set (6.5), makes it possible to find the trivial $\varepsilon_0\sim\alpha_1$ and

$$\lambda_0=-\frac{\varkappa_2\alpha_1}{8\varkappa_1}\left[-1+\sqrt{1+\frac{8\gamma\varkappa_1}{\beta_1\varkappa_2}}\right]^2-\alpha_2+\frac{\gamma\alpha_1}{\beta_1}. \tag{6.6}$$

Equating $\lambda_0=0$ enables us to find the boundary of stability of the solution (6.3) for the domain boundary in the phase diagram $(\alpha_1,\alpha_2)$ [69]. According to (6.6), the equation of line 3 in Fig. 2.10 is a straight line

$$-\alpha_1=\frac{-8(\varkappa_1/\varkappa_2)\alpha_2}{q-\left[1-\sqrt{1+q}\right]^2}, \quad q=\frac{8\varkappa_1}{\varkappa_2}\frac{\gamma}{\beta_1}. \tag{6.7}$$

Distribution in the domain wall with $\eta\neq 0$ and $\varphi\neq 0$ is evidently implemented to the left of line 3. The appearance of a new non-zero parameter means that here in the vicinity of the original domain wall a layer of a new phase ($\varphi \neq 0$) appears, and it is localized close to the domain wall. The localization of the mentioned layer is confirmed by the well-known form of the eigenfunction $\psi_0$ of the second equation of set (6.5).

When discussing the phase transformation in the domain wall in the material with multicomponent order parameters ($\eta_1$, $\eta_2$) it should be mentioned that from the formal viewpoint the difference in describing the phase transition in comparison with the previous material consists of the existence here of an additional link between the expansion coefficients in (6.1): $\alpha_1 = \alpha_2$, $\beta_1 = \beta_2$, $\varkappa_1 = \varkappa_2$. Although the plan of consideration itself of the phase transition in the material with several order parameters and in the material with the multicomponent parameter is the same, the interpretation of the obtained results in the latter case changes qualitatively. The appearance of the additional component $\eta_2 \neq 0$ in the domain wall as the result of the phase transition in comparison with for example the initial single-component distribution indicates here the rotation of the order parameter and not the precipitation of a new phase. As a result instead of the flat domain boundary (the order parameter is everywhere in a single plane) appears a domain boundary of the rotational type (Fig. 2.11) [62]. And of course, at the same time, the rotation in the boundary may also be accompanied by a change of the modulus of the order parameter. It is reasonable to assume that such phase transition in the domain wall takes place in the antiphase domain boundary of the improper ferroelectric crystal of gadolinium molybdate [70].

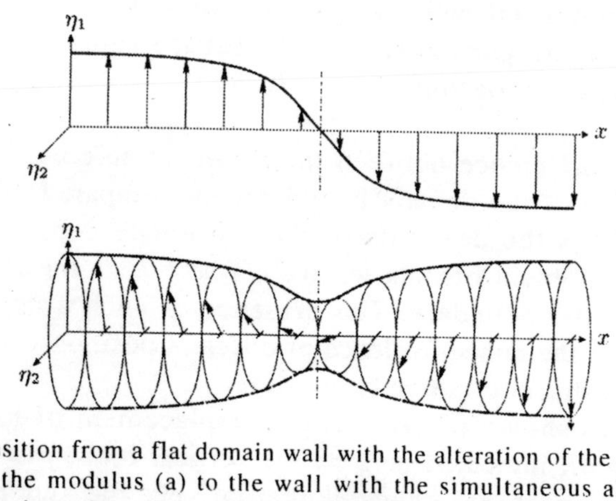


**Fig.2.11.** Transition from a flat domain wall with the alteration of the order parameter in respect of the modulus (a) to the wall with the simultaneous alteration of the modulus and orientation of the order parameter (b).

Chapter 3

# Discussion of the microscopic structure of the domain boundaries in ferroelectrics

The results of Chapter 2 show that with the exception of a range in the vicinity of Curie point $T_c$, the domain boundaries in ferroelectrics always remain narrow with the width of only several constants of the elementary cell. It is evident that for such boundaries the application of the results of continuous consideration where the functional dependence (for example, dependence (1.27) and (1.26) in Chapter 2), describing the alternation of the polarization vector in the boundary, can in the limit simply fall on the atomic gap is not acceptable for serious numerical estimates. In this case, it is necessary to carry out microscopic investigation taking into account details of a specific structure and particularities of interactions in a specific material.

## 3.1. Lattice potential relief for a domain wall Calculation of the parameters of the lattice relief in quasi-continuous approximation

The principal difference between the results of microscopic analysis of the structure of the domain boundaries as compared to continuous approximation is the detection of the coordinate dependence of the energy of the domain boundaries $\gamma(U)$, where $U$ is the coordinate of the centre of the boundary. The presence of such a dependence is associated with the nonequivalence of different positions of the domain boundary in a discrete lattice (Fig. 3.1).

Actually, as shown in Fig.3.1, the displacement of the centre of the domain wall, indicated here by the vertical broken line, over half of the atomic space $a/2$ changes qualitatively the symmetry of the relative distribution of the atoms and of the associated dipole

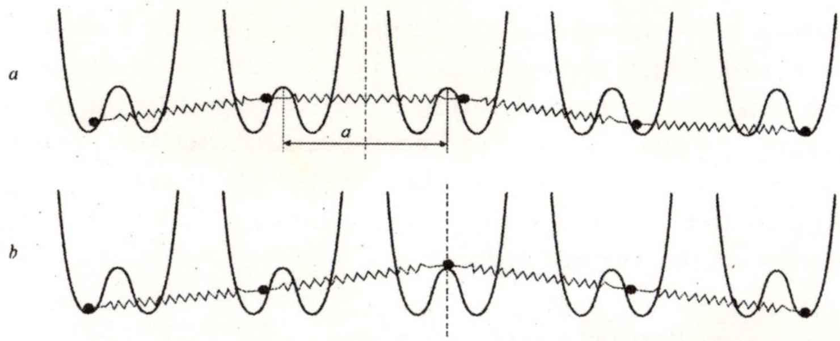


**Fig. 3.1.** Distribution of the displacement of ferroactive particles in different configurations of a narrow domain wall in ferroelectrics.

moments in the boundary. Therefore, the energy of the corresponding configurations of the domain wall may also differ (Fig. 3.2) and this results in a periodic lattice relief for the domain wall similar to the Peierls relief for dislocations. The difference in the energy of the domain wall configurations shown in Fig. 3.1 is usually equal to the value of the lattice energy barrier $V_0$.

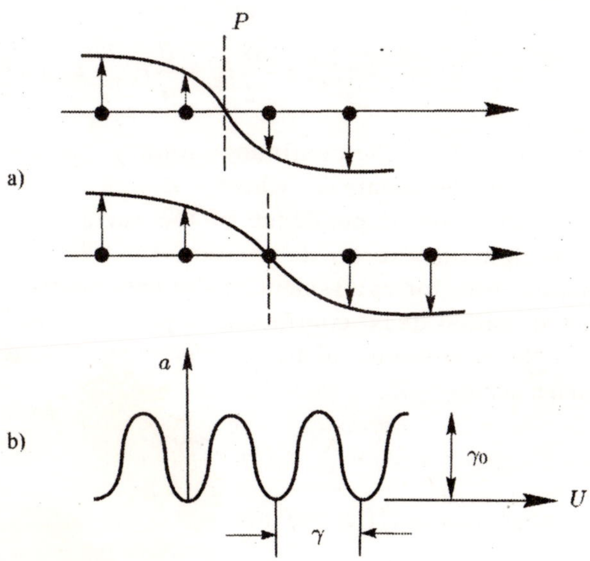


**Fig. 3.2.** Alternation of the polarization vector for different configurations of the narrow domain wall (a). The periodic dependence of the surface density of the domain wall energy taking into account the discreteness of the crystal lattice (b): $V_0$ is the lattice energy barrier for the domain wall.

Let us consider the alternation of the surface energy of the domain wall, which accompanies its uniform motion in the crystalline lattice. Taking into consideration the discreteness of the crystalline structure results in the replacement of the integral in the thermodynamic potential of the crystal by a sum of members related to individual atomic planes. In a crystal with the phase transition of the second order

$$\Phi = a\sum_{n=-\infty}^{\infty}\left\{-\frac{\alpha}{2}P_n^2+\frac{\beta}{4}P_n^4+\frac{\varkappa}{2a^2}\left(P_n-P_{n+1}\right)^2\right\}. \quad (1.1)$$

Here $n$ is the number of the atomic plane, $a$ is the interplanar spacing, $P_n = P(na)$ is the value of the polarization vector in the $n$-th atomic plane. Extremes (1.1) correspond to the values of $P_n$, which are solutions of the difference equation

$$\frac{\varkappa^2}{a}\left(P_{n+1}-2P_n+P_{n-1}\right)=-\alpha P_n+\beta P_n^3 \quad (1.2)$$

with the boundary conditions $P_n \to \pm P_0$ at $n \to \pm\infty$. Substitution of (1.2) into (1.1), less the energy of the homogeneous state, gives

$$\gamma = a\sum_n\left\{\varphi(P_n)-\varphi(P_0)-\frac{1}{2}P_n\frac{d\varphi}{dP_n}\right\}=a\sum_{n=-\infty}^{\infty}f(na-U), \quad (1.3)$$

where, as previously, $\varphi(P_n)=-\frac{\alpha}{2}P_n^2+\frac{\beta}{4}P_n^4$, and $P_n$ satisfies equation (1.2).

Transition in (1.3) to the continuous limit yields quantity $\gamma$ that is independent of coordinate $U$, whereas direct calculations of the lattice sum lead to the dependence of the surface density of the energy of the domain wall on its position $\gamma = \gamma(U)$.

The simplest way for assessment of the parameters of the lattice relief is the so-called quasi-continuous approximation [71]. Taking into account the periodicity of the dependence $\gamma(U)$, we expand it into a Fourier series

$$\gamma(U)=\sum_{m=-\infty}^{\infty}\gamma(U)\cdot e^{2\pi imU/a}, \quad (1.4)$$

where the expansion coefficients are

$$\gamma_m=\frac{1}{a}\int_0^a\gamma(U)\cdot e^{-2\pi imU/a}dU=\sum_n\int_0^a f(na-U)\cdot e^{-2\pi imU/a}dU. \quad (1.5)$$

Multiplying additionally (1.5) by the identical unit $e^{2\pi imn}$ and substituting variable $U-na\equiv U'$, instead of (1.5) we obtain

$$\gamma_m = \sum_n \int_{-na}^{(-n+1)a} f(U') \cdot e^{-2\pi i m U'/a} dU' = \int_{-\infty}^{\infty} f(x) \cdot e^{-2\pi i m x/a} dx. \tag{1.6}$$

The main approximation in the framework of the quasi-continuous approach is the application in calculation (1.6) not of the real dependence of $P_n$ on $n$ from (1.2) but of the continuous dependence $P(x) = P_0$ th$(x/\delta)$, found in Chapter 2. In this case, taking into account the explicit form

$$f(x) = \varphi(P) - \varphi(P_0) + \frac{\varkappa}{2}\left(\frac{dP}{dx}\right)^2 = \frac{\beta}{2} P_0^4 \cdot \frac{1}{\mathrm{ch}^4 x/\delta} \tag{1.7}$$

(see section 2.1) the coefficients of the expansion (1.6) are:

$$\begin{aligned} \gamma_0 &= \frac{\beta}{2} P_0^4 \delta \int_{-\infty}^{\infty} \frac{dx/\delta}{\mathrm{ch}^4 x/\delta} = \frac{2}{3} \alpha P_0^2 \delta, \\ \gamma_{|m|>0} &= \frac{\beta}{2} P_0^4 \delta \int_{-\infty}^{\infty} \frac{\cos(2\pi m x/a)}{\mathrm{ch}^4 x/\delta} dx/\delta = \\ &= \beta P_0^4 \delta \cdot \frac{2}{3} \pi^2 m \left(\frac{\delta}{a}\right) \frac{1}{\mathrm{sh}(\pi^2 m \delta/a)} \cdot \left[\frac{\pi^2 m^2 \delta^2}{a} + 1\right]. \end{aligned} \tag{1.8}$$

The strong decrease of the coefficients $\gamma_{|m|>0}$ with number $m$ enables us to retain in equation (1.4) one member of the sum, besides the zero number, that results in the dependence $\gamma(U)$:

$$\begin{aligned} \gamma(U) &= \gamma_0 + \frac{V_0}{2} \cos \frac{2\pi n U}{a}, \\ V_0 &= 8\pi^4 \gamma_0 \left(\frac{\delta}{a}\right)^3 \cdot e^{-\pi^2 \delta/a}. \end{aligned} \tag{1.9}$$

According to (1.9), the dependence $\gamma(U)$ is periodic in fact with the value of the barrier $V_0$, which strongly depends on the relative width of the domain wall (in comparison with the lattice constant). At $\delta \gg a$, as in the case with, for example, ferromagnetics and ferroelectrics in the vicinity of the Curie point, the barrier in the dependence $\gamma(U)$ almost completely disappears, whereas for narrow domain boundaries with $\delta \sim a$ its presence as shown below has a strong influence on the possibility of displacement of the domain boundaries.

## 3.2. Calculation of electric fields in periodic dipole structures. Determination of the correlation constant in the framework of the dipole-dipole interaction

The research of the microstructure of domain boundaries in some materials is based on the determination of electric fields in the area where structural units are located, as well as on the calculation of the energy of the dipole–dipole interaction, which plays an especially significant role in ferroelectrics. Below we present a scheme, which enables rather quick calculation of the electric fields in various structures using an universal method.

As it is known from electrostatics, the dipole contribution to the electrostatic potential $\varphi_d$, obtained as a result of expansion of the potential of the system of charges into a series, has the form

$$\varphi_d(\mathbf{r}) = -\sum_j P_j \frac{\partial}{\partial x_j} \cdot \frac{1}{|\mathbf{r}-\mathbf{r}'|}, \tag{2.1}$$

where $P_j$ is the $j$-th projection of the dipole moment of the system located in the point with the coordinates $\mathbf{r}'$. Consequently, the field created by the dipole moment is

$$E_i = -\frac{\partial \varphi_d}{\partial x_i} = \sum_j P_j \frac{\partial^2}{\partial x_i \partial x_j} \cdot \frac{1}{|\mathbf{r}-\mathbf{r}'|} \equiv \sum_j I_{ij} . P_j, \tag{2.2}$$

where

$$I_{ij} = \frac{\partial^2}{\partial x_i \partial x_j} \cdot \frac{1}{|\mathbf{r}-\mathbf{r}'|}, \tag{2.3}$$

is the so-called structural factor.

In specific structural calculations we have to separate the sub-lattices of different dipoles. Giving them indices $\mu$ and $\nu$, and also taking into account the symmetry of the problem, i.e. the fact that all dipole moments of the $\nu$-th type $P_{\nu j}(x'_\nu)$ at fixed coordinate $x'_\nu$ are identical along the direction of the normal to the boundary and having taken the above into consideration introducing the structural factor for the dipole plane

$$I_{ij}\left(x_\mu - x'_\nu\right) = \sum_{y'_\nu, z'_\nu} \frac{\partial^2}{\partial x_{\mu_i} \partial x_{\mu_j}} \cdot \frac{1}{|\mathbf{r}_\mu - \mathbf{r}'_\nu|}, \tag{2.4}$$

the $i$-th component of the electric field in the location of the $\mu$-th type dipole can be presented in the form

$$E_{\mu i} = \sum_{x'_\nu} \sum_{\nu} \sum_{j} I_{ij}\left(x_\mu - x'_\nu\right) \cdot P_{\nu j}\left(x'_\nu\right). \tag{2.5}$$

In equation (2.5) summation is carried out in respect of the dipole planes correspondingly, types of dipoles or dipole complexes and projections of dipole moments. The prime in the sum (2.4) indicates that the self-action of the corresponding dipole is not taken into account.

Slow convergence in summation in (2.4) inhibits calculations in the real space in this case. It is far more convenient to carry out calculations using Fourier expansion. For this purpose, let us in the first place go over from $\mathbf{r}_\mu$ and $\mathbf{r}'_\nu$ to new variables $\mathbf{r}$ and $\mathbf{r}_{\mu\nu}$, where vector $\mathbf{r}$ describes the position, of an elementary cell in the crystal and vector $\mathbf{r}_{\mu\nu}$ is the relative distribution of the dipole units within the limits of the elementary cell. Then, instead of the difference in (2.4) $\mathbf{r}_\mu - \mathbf{r}'_\nu$ we have $\mathbf{r} - \mathbf{r}'_{\mu\nu} = \mathbf{r}_\mu - \mathbf{r}'_\nu$. Going over further to the dimensionless variables $n_i = x_i/a_i$ and $s_{i\mu\nu} = x_{i\mu\nu}/a_i$, where $x_i$ and $x_{i\mu\nu}$, are the components of vectors $\mathbf{r}$ and $\mathbf{r}_{\mu\nu}$ respectively, and $a_i$ are the dimensions of the elementary cell along the corresponding axes, let us introduce a dimensionless structural (lattice) factor instead of (2.4), which for a lattice close to the cubic one has the form:

$$I_{ij}\left(n_1, \mathbf{s}\right) = V_{\text{cell}} \cdot I_{ij}\left(x_\mu - x'_\nu\right) = \sum_{n_2, n_3} \frac{\partial^2}{\partial s_i \partial s_j} \cdot \frac{1}{|\mathbf{n} - \mathbf{s}|}. \tag{2.6}$$

Here $\mathbf{n} = (n_1, n_2, n_3), \mathbf{s} \equiv \mathbf{s}_{\mu\nu}$, $V_{\text{cell}}$ is the volume of the elementary cell.

Let us expand sum $\sum \frac{1}{|\mathbf{n} - \mathbf{s}|}$ in (2.6) into a Fourier series. Then, instead of (2.6) we can write

$$I_{ij}\left(n_1, \mathbf{s}\right) = \frac{\partial^2}{\partial s_i \partial s_j} \sum_{m_2, m_3} I\left(m_2, m_3\right) \cdot \exp\left(-2\pi i\left(m_2 s_2 + m_3 s_3\right)\right), \tag{2.7}$$

where the coefficient of Fourier expansion is

$$I\left(m_2, m_3\right) = \int_0^1 \int_0^1 \left(\sum_{n_2, n_3} \frac{1}{|\mathbf{n} - \mathbf{s}|}\right) \cdot \exp\left(-2\pi i\left(m_2 s_2 + m_3 s_3\right)\right), \tag{2.8}$$

To calculate (2.8) we use the following transformation. The right hand part of (2.8) is multiplied by the value exp $(2\pi i(m_2 n_2 + m_3 n_3))$ identical to 1 and instead of $\mathbf{s}$ let us introduce a new vector $\mathbf{s}' = (s_1, s_2 - n_2, s_3 - n_3)$. Then, in new variables

$$I(m_2,m_3)=\sum_{n_2,n_3}\int_{-n_2}^{-n_2+1}\int_{-n_3}^{-n_3+1}\frac{1}{\sqrt{s_2'^2+s_3'^2+(n_1-s_1')^2}}\times$$
$$\times\exp\left(-2\pi i\left(m_2s_2'+m_3s_3'\right)\right)\cdot ds_2'ds_3'. \tag{2.9}$$

Replacing the sum of the individual integrals by integrals in infinite limits, we obtain

$$I(m_2,m_3)=\int_{-\infty}^{\infty}\int_{-\infty}^{\infty}\frac{1}{\sqrt{s_2'^2+s_3'^2+(n_1-s_1')^2}}\times$$
$$\times\exp\left(-2\pi i\left(m_2s_2'+m_3s_3'\right)\right)\cdot ds_2'ds_3', \tag{2.10}$$

Transferring to the polar system of coordinates $s_2'=\rho\cos\varphi$, $s_3'=\rho\sin\varphi$ we have $s_2'^2+s_3'^2=\rho^2$, $s_2'^2+s_3'^2=m\rho\cos\varphi$ and consequently

$$I(m_2,m_3)=\int_0^{2\pi}\int_0^{\infty}\frac{1}{\sqrt{\rho^2+d^2}}\cdot\exp(-2\pi i m\rho\cos\varphi)\cdot\rho d\rho d\varphi=$$
$$=\int_0^{\infty}2\pi\frac{I_0(2\pi m\rho)}{\sqrt{\rho^2+d^2}}\cdot\rho d\rho=\frac{2\pi\exp(-2\pi m\alpha)}{\alpha}, \tag{2.11}$$
$$\alpha=|n_1-s_1|.$$

Returning to (2.7) taking into account the symmetric form of the dependence of (2.11) on $m_2$, $m_3$ $\left(m=\sqrt{m_2^2+m_3^2}\right)$, we finally have the following equation for the structural factor of the dipole plane [72]:

$$I_{ij}(\vec{s},n_1)=\frac{\partial^2}{\partial s_i\partial s_j}\sum_{m_2,m_3}2\pi\frac{1}{|n_1-s_1|}\times$$
$$\times\exp\left\{-2\pi|n_1-s_1|\sqrt{m_2^2+m_3^2}\right\}\times \tag{2.12}$$
$$\times\cos(2\pi m_2s_2)\cos(2\pi m_3s_3).$$

The ratio (2.12) is suitable only for materials with a small deviation from the cubic shape of the initial cell at the transition of the material to the ferroelectric phase. In other cases, instead of (2.12), the lattice factor for the corresponding plane can be obtained in the form of the sum of the lattice factors for dipole lines. The same procedure should be used to calculate the lattice factor in any case for the so-called 'own' plane, i.e. the plane passing through the dipole, in the location of which electric field is sought. As expected, in this case, i.e. at $n_1=s_1$, a divergence appears in equation (2.12).

The calculation of the structural factor for the dipole line is carried out using the same procedure as in the case of the dipole plane, except that in this case we use expansion into a one-dimensional Fourier integral. For the line, oriented along axis $z$

$$I_{ij}(n_1,n_2,\mathbf{s})=\sum_{n_3}\frac{\partial^2}{\partial s_i\partial s_j}\cdot\frac{1}{|n_1-s_1|}= \\ =\frac{\partial^2}{\partial s_i\partial s_j}\sum_{m_3}I(m_3)\exp(2\pi i m_3 s_3), \tag{2.13}$$

$$I(m_3)=\int_0^1\left(\sum_{n_3}\frac{1}{|n_1-s_1|}\right)\cdot\exp(-2\pi i m_3 s_3)ds_3= \\ =\int_{-\infty}^{\infty}\frac{1}{\sqrt{(n_1-s_1)^2+(n_2-s_2)^2+s_3^2}}\times \\ \times\exp(-2\pi i m_3 s_3')ds_3'. \tag{2.14}$$

After changing notation $(n_1-s_1)^2+(n_2-s_2)^2=\alpha^2$ we obtain

$$I(m_3)=\int_{-\infty}^{\infty}\frac{1}{\sqrt{\alpha^2+s_3^2}}\cdot\exp(-2\pi i m_3 s_3')ds_3'= \\ =2K_0\left(2\pi m_3\sqrt{(n_1-s_1)^2+(n_2-s_2)^2}\right), \tag{2.15}$$

where $K_0(x)$ is the cylindrical Macdonald function.

Taking into account (2.15) and the possible difference in the parameters of the elementary cell, which is not taken into account in (2.15), the structural factor for 'foreign' dipole line has the form

$$I_{ij}(\mathbf{s},n_1,n_2)=\frac{\partial^2}{\partial s_i\partial s_j}\cdot 2\pi\sum_{m_3}\frac{1}{\pi}\times \\ \times K_0\left(2\pi\frac{m_3}{a_3}\sqrt{a_1^2(n_1-s_1)^2+a_2^2(n_2-s_2)^2}\right)\cdot\cos(2\pi m_3 s_3). \tag{2.16}$$

The lattice factor for 'own' line, i.e. the line passing through the point, at which the following field is sought

$$I = 2\sum_{n_3=-\infty}^{\infty}{}' \cdot \frac{1}{\left|n_3 - s_3\right|^3}. \tag{2.17}$$

The equations obtained above make it possible to estimate in the first place the value of the correlation constant used in Chapter 2, assuming that the main contribution comes from the dipole–dipole interaction. For this purpose, let us consider the simplest cubic lattice of the dipoles oriented along the single axis $z$ and let us assume that the alteration of their values depends only on single coordinate $x$. That enables us to remove all the indices in equation (2.5) and to write the volume density of the energy of the dipole–dipole interaction in the form of the sum

$$\Phi_n = \sum_m -I(n-m) \cdot P_n P_m \tag{2.18}$$

Here $n$, $m$ are the numbers of atomic planes, $P_n$ and $P_m$ are the values of the polarization vector in these planes, $I(n-m)$ is the dimensionless structural factor (2.6), which depends on the difference between the numbers of the planes.

Taking into account the heterogeneity of the distribution of the polarization vector its values in the planes adjacent to the $n$-th plane can be written in the form of expansion

$$\begin{aligned} P_{n+1} &= P_n + \frac{dP_n}{dx}a + \frac{1}{2}\frac{d^2P_n}{dx^2}a^2 + \ldots, \\ P_{n-1} &= P_n - \frac{dP_n}{dx}\cdot a + \frac{1}{2}\frac{d^2P_n}{dx^2}a^2 + \ldots \end{aligned} \tag{2.19}$$

According to (2.12), the lattice factor $I(n-m)$ decreases exponentially with the increase of the argument, which allows us to retain in (2.18) the interaction only with the dipoles of the 'own' and adjacent atomic planes. As the result, expression (2.18) taking into account the cancellation of the number of the terms containing the first derivatives, can be written in the form of

$$\Phi_n = -\left[I(0) + 2I(1)\right] \cdot P_n^2 - \frac{1}{2}a^2 \cdot I(1) \cdot P_n \cdot \frac{d^2P_n}{dx^2} \tag{2.20}$$

or, after calculating sum $\sum_n \Phi_n \to \int \Phi(x)dx/a$ using the results of integration by parts, in the form of

$$\Phi_n = \Phi_{n0} + \frac{1}{2}a^2 \cdot 2I(1) \cdot \left(\frac{dP_n}{dx}\right)^2,$$
$$\Phi_{n0} = -\left[I(0) + 2I(1)\right] \cdot P_n^2. \tag{2.21}$$

As indicated in (2.21), the second term in $\Phi_n$ coincides here with the ordinary correlation term in (1.1) at

$$\varkappa = a^2 \cdot 2I(1). \tag{2.22}$$

For a simple cubic lattice from (2.16)

$$I(1) = -(2\pi)^3 \sum_{m_2, m_3} m_3^2 \cdot \exp\left(-2\pi\sqrt{m_2^2 + m_3^2}\right) \simeq -0{,}926. \tag{2.23}$$

Thus, here $|\varkappa| \simeq 1.85a^2$ or $\sim a^2$ as it is usually assumed when making estimates on the basis of the order of magnitude.

### 3.3 Structure of the 180° and 90° domain walls in barium titanate crystal

Microscopic calculations of the structure of the domain boundaries are based on the structure of a specific material. The elementary cell of one of the few crystals, for which such calculations have been carried out (the crystal of barium titanate $BaTiO_3$) is shown in Fig.3.3

The structure of barium titanate is based on oxygen octahedrons $TiO_6$ centred by titanium and linked via their vertexes. Barium cations are located in the spaces between the octahedrons.

In the initial paraelectric phase the elementary cell of $BaTiO_3$ has a cubic symmetry with lattice parameter equal to approximately 4 Å. At $T$ = 120°C the crystal changes to a tetragonal ferroelectric phase, which transforms to a rhombic phase at 0 °C, and then to a rhombohedral phase at –90÷70 °C.

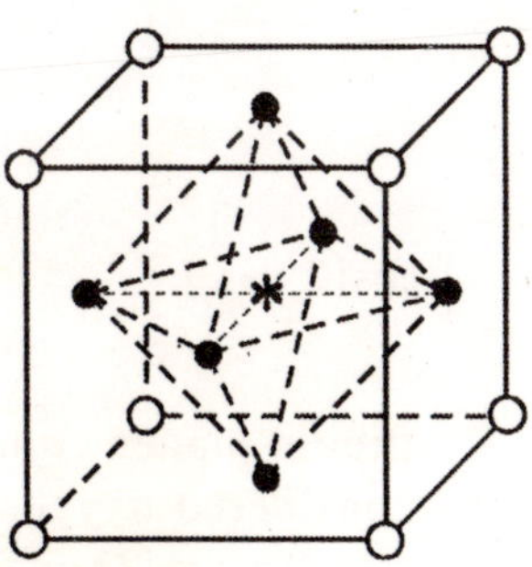

**Fig. 3.3.** The elementary cell of barium titanate: o – barium, • – oxygen,* – titanium [2].

Spontaneous polarization in the corresponding ferroelectric phases is firstly directed along the rib, then along the diagonal of the edge and finally along the spatial diagonal of the cube, respectively.

For further calculations, let us numerate the ions of different type in the elementary cell. Let us assign values of the indices $\mu, \nu$ =

0,1,2,3,4 to the ions of Ti, $O^1$, $O^2$, $O^3$ and Ba respectively, and at the same time let us assign the values $\mu, \nu = 2$ to the oxygen ions that are located in the same plane with titanium ions, which is parallel to the domain boundary. To determine the structure of the domain boundaries, we use Slater's model in which it is assumed that ion polarization $P'$ is produced only by displacement of the titanium ion, and all ions have electronic polarizabilities. The dipole moments, created by any displacement of the charges, are assumed to be of the point type and located in the areas occupied by the corresponding ions.

Let us examine the structure of the 180° domain wall. To determine the distribution of the total polarization in the boundary, it is necessary to find the ratio between the ion and electron polarization. When calculating the electronic polarization, let us confine ourselves to the largest contribution to polarization $\tilde{P}$, which will be obviously provided by the electron displacement in the oxygen ion $O^1$ located on the same straight line with the titanium ions, parallel to the polarization vector. The mentioned ratio between $\tilde{P}$ and $P'$ in the arbitrary $n$-th atomic plane in the boundary, containing titanium ions and the mentioned oxygen ions $O^1$ can be found by calculating the strength of the electric field generated in the location of the oxygen ion $O^1$.

Assuming that the originating polarization is directed along axis $z$, according to the expressions (2.16) and (2.7) of the previous section, the strength of the electric field in the location of oxygen $O^1$ in the $n$-th atomic plane can be written in the form of

$$E_n\left(O^1\right) = \sum_{n'} \left\{ I_0\left(n-n'\right) \cdot \tilde{P}_{n'} + I_{1/2}\left(n-n'\right) \cdot P'_{n'} \right\}, \qquad (3.1)$$

where

$$\begin{aligned} I_{s_3}\left(n-n'\right) = \delta_{n-n',0} \sum_{n_3}{}' \frac{2}{\left|n_3 - s_3\right|^3} - \\ -(2\pi)^3 \sum_{n_2} \left(1 - \delta_{n-n',0} \cdot \delta_{n_2,0}\right) \times \\ \times \sum_{m=1}^{\infty} m^2 \frac{2}{\pi} K_0\left(2\pi m \sqrt{2\pi m \left(n_1 - n'\right)^2 + n^2}\right) \cdot \cos\left(2\pi m s_3\right). \end{aligned} \qquad (3.2)$$

Here, $I_0$ and $I_{1/2}$ are structural factors for dipole planes compiled from dipoles situated in the locations of the considered oxygen ions $O^1$ and titanium ions, respectively. The indices 0 and 1/2 of these factors correspond to the value $s_3$, which should be used when calculating the corresponding factors on the basis of equation (3.2).

Due to the rapid decrease of the value of the structural factors $I_0(n-n')$ and $I_{1/2}(n-n')$ with increasing argument let us confine ourselves to several terms in the sum (3.1). Introducing the electron polarizability of the oxygen ions $\alpha$ in the form of the ratio $\tilde{P} = \alpha \cdot E(O^1)$, on the basis of (3.1) we obtain the following equation, linking ion and electron polarization:

$$\begin{aligned}&(\alpha_{-1} - \mathcal{I}_0)\cdot \tilde{P}_n - I_0(1)\cdot(\tilde{P}_{n+1} - 2\tilde{P}_n - 2\tilde{P}_{n-1}) = \\ &= \mathcal{I}_{1/2}\cdot P_n' - I_{1/2}(1)\cdot(\tilde{P}_{n+1} - 2P_n' + P_{n-1}').\end{aligned} \tag{3.3}$$

Here $\mathcal{I}_0 = I_0(0) - 2I_0(1)$, $\mathcal{I}_{1/2} = I_{1/2}(0) - 2I_{1/2}(1)$. The following numerical values of the lattice sums are obtained: $I_0(0) = 4.5$, $I_{1/2}(0) = 32.5$, $I_0(1) = -0.1$, $I_{1/2}(1) = 0.1$.

In the limit of the narrow wall, where $P_{n-1}=P_n=-P_{n+1}$, the solution of equation (3.3) can be expressed with the help of $P_n=P'_n+\tilde{P}_n$, which taking into account specific values of the lattice factors and polarizability $a = 3.7\cdot 10^{-2}$ [2] yields here

$$\tilde{P}_n = 0.6\cdot P_n, \quad P_n' = 0.4\cdot P_n. \tag{3.4}$$

When the ratio between $\tilde{P}_n$ and $P'_n$ is available, we can write the equation for the distribution of total polarization in the boundary. As previously, let us use the value of the strength of the electric field calculated on the basis of (3.1) and (3.2) this time in the area where the titanium ion is located. According to (3.1) and (3.2)

$$\begin{aligned}E(\mathrm{Ti}) = &\mathcal{I}_0 P_n' + I_0(1)\cdot(P_{n+1}' - 2P_n' + P_{n-1}') + \\ &+\mathcal{I}_{1/2}\cdot\tilde{P}_n + I_{1/2}(1)\cdot(\tilde{P}_{n+1} - 2\tilde{P}_n + \tilde{P}_{n-1}).\end{aligned} \tag{3.5}$$

On the other hand, the same strength can be found as a derivative of thermodynamic potential $\partial\Phi/\partial P'_n = -E_n$

$$\Phi = \sum_n \left\{ -\frac{\alpha}{2}P_n'^2 + \frac{\beta}{4}P_n'^4 + \frac{\gamma}{6}P_n'^6 \right\}, \tag{3.6}$$

where $\alpha$, $\beta$, $\gamma$ are determined by short-range atomic potentials. Equating the above expressions and taking (3.4) into account gives the following equilibrium equation describing the distribution of the total polarization in the boundary:

$$\begin{aligned}&\frac{\varkappa}{a^2}(P_{n+1} - 2P_n + P_{n-1}) = -\alpha P_n + \beta P_n^3 + \gamma P_n^5, \\ &\varkappa = \left[0.4 I_0(1) + 0.6 I_{1/2}(1)\right]\cdot a^2 \simeq 0.02\cdot a^2.\end{aligned} \tag{3.7}$$

As expected, equation (3.7) is completely identical with the general equation (2.2), the only difference being that here we have found the structure of different contributions to the coefficients of expansion of the thermodynamic potential.

The solution of the set of equations (3.7) for different boundary configurations (Fig. 3.1, *a*, *b*) can be found by the method used when considering the structure of the dislocation nucleus. The symmetry of distribution of the values of the polarization vector for different atomic planes in two configurations of the domain wall with the extreme values of energy is described by the equations of the type:

$$\begin{aligned} &I.\quad P_{n=0}=0, \qquad P_n=-P_{-n,} \quad n=\pm1,\pm2,\dots \\ &II.\quad P_{-n+1}=-P_{n,} \quad n=\pm1,\pm2,\dots \end{aligned} \tag{3.8}$$

In method [73] the values of the polarization vector in the centre of the boundary (planes with the numbers $n$ = 0, ±1 in configuration I and planes with numbers $n$ = 0, −1 in configuration II) are determined directly from the simultaneous equations (3.7). For the remaining part of the boundary, the alternation of polarization with number $|n|$ is simulated by the dependence on the type

$$P_n=P_0\cdot\left[1-c\cdot\exp\left(-|n|\cdot\lambda\right)\right], \quad n\geq 2, \tag{3.9}$$

where parameter $\lambda$, characterising the width of the boundary, is determined from this self-consistent condition obtained as a result of substitution of the approximate solution (3.9) into exact equation (3.7) at the limit of high $n$ values. Coefficient $c$ in (3.9) is determined from the conditions of joining the solutions (3.7) and (3.9) in the centre of the boundary at $n$ = 1.

The calculations carried out in [72] for the values obtained here at $T$ = 20°C, $\alpha$ = 30.2·10$^{-3}$, $\beta$ = 0.9·10$^{-12}$, $\gamma$ = 54·10$^{-23}$ and $P_0$=8·10$^4$ CGSE units, revealed the following distribution of polarization in configurations I and II:

$$I.\ P_n=\begin{cases} 0, & n=0 \\ 0.8\cdot P_{0,} & |n|=1 \\ P_0, & |n|>1 \end{cases} \quad II.\ P_n=\begin{cases} 0.6P_{0,} & |n|=1 \\ 0.9P_0, & |n|=2 \\ P_0, & |n|>2 \end{cases} \tag{3.10}$$

Calculations of the energies for the given boundary configuration show that at room temperature configuration II is stable or basic. The value of $\gamma_0$ for it obtained from equation (1.3) by adding here the term $\frac{\gamma_0}{6}\left(P_n^6-P_0^6\right)$ turns out to be equal to 6.3 erg·cm$^{-2}$. On the other hand, configuration I is a saddle or barrier configuration with the value of $\gamma = \gamma_0 + V = 7$ erg·cm$^{-2}$. Thus, the value of the lattice barrier that the given wall has to overcome in its motion is 0.7 erg·cm$^{-2}$, which is close to the data in [74,75].

Identical calculations, carried out in [76,77] for the 90° domain wall in $BaTiO_3$, showed that in the first approximation its structure corresponds to the conclusions of continuous approximation. In this case, the width of the domain wall is $\delta \simeq 5.2a \simeq 21$ Å and the surface density of its energy is $\gamma_0 \simeq 7.1$ erg/cm$^2$. The lattice energy barrier for this wall is negligible. Like in the phenomenological consideration [55], the numerical calculations of the 90° domain wall structure in $BaTiO_3$ confirm the presence of heterogeneity in the distribution of the polarization component normal to the plane of the boundary leading to the formation of an internal electric field in such a boundary.

## 3.4 Structure of the domain boundaries in ferroelectric crystals of the potassium dihydrophosphate group

The ferroelectric crystal of potassium dihydrophosphate $KH_2PO_4$ (KDP) has the tetragonal symmetry in the initial paraelectric phase and the orthorhombic symmetry below the Curie point $T_c = 123$ K. The crystalline lattice of this compound consists of two body-centred sublattices of $PO_4$ inserted in each other and two body-centred sublattices of K, and for all that the lattices of $PO_4$ and K are displaced along the polar $z$-direction (Fig. 3.4). $PO_4$ tetrahedrons are connected by hydrogen bonds that are almost normal to the ferroelectric axis.

The transition to the ferroelectric phase is accompanied by the ordering of protons on hydrogen bonds. The value of polarization observed away from $T_c$ $P_0 = 5.1$ μC·cm$^{-2}$ [2] is explained by the displacement of $K^+$, $P^{5+}$, $O^{2-}$ ions along the $z$ axis in relation to their symmetric positions. The displacements of the protons themselves on the hydrogen bonds do not provide almost any contribution to the value of $P_0$, but it is assumed that their ordering is the reason for the displacement of the remaining ions, which provide a direct contribution to spontaneous polarization.

The crystal of potassium dihydrophosphate has many compounds isomorphous to itself, which are also ferroelectrics. They are formed by means of substitution of K → Rb, Cs and P → As, and also of hydrogen by deuterium H → D.

In the cluster approximation, the $KH_2PO_4$ crystal is represented

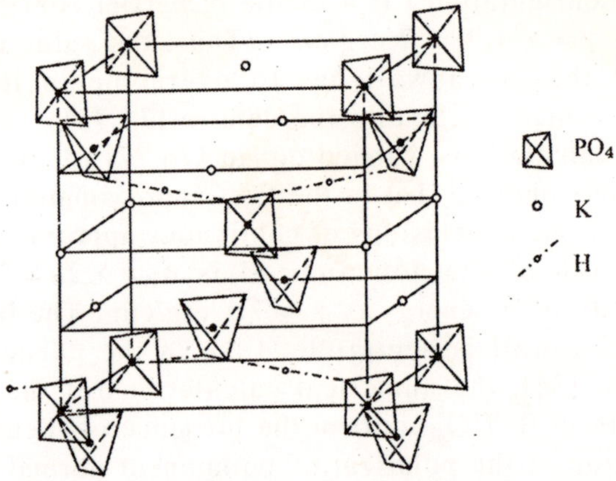


**Fig. 3.4.** The structure of an elementary cell of the $KH_2PO_4$ crystal according to West [2].

[6] in the form of a set of configurations (Fig. 3.5) with a different number of protons adjacent to the $PO_4$ tetrahedron. Among these configurations, there are polar and neutral configurations, each of them has a specific energy and, consequently, a specific probability of being implemented.

At low temperatures in the ferroelectric phase the volume of each domain may be regarded as consisting of specific configurations of type 1. According to [78] in this case (Fig. 3.6) the domain boundary represents a monomolecular layer of type 2 configurations. Due to the symmetry of distribution of protons near them, the configurations of type 2 can also be assigned a specific dipole moment now oriented in the direction normal to the vector $\mathbf{P}_0$. As it could be seen from Fig.3.6, two types of domain boundaries can be formed consisting of type 2 configurations. The

| | Configuration | Energy | Dipole moment | Formula |
|---|---|---|---|---|
| 1 | 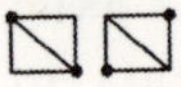 | 0 | $\pm P$ | $H_2PO_4$ |
| 2 | 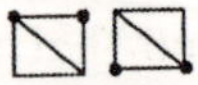 | $\mathcal{E}_0$ | 0 | $H_2PO_4$ |
| 3 | 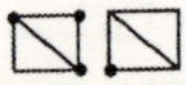 | $W$ | $\pm P/2$ | $H_3PO_4$, $HPO_4$ |
| 4 |  | $W_1$ | 0 | $H_4PO_4$, $PO_4$ |

**Fig. 3.5.** Schematic image of the $PO_4$ tetrahedrons in the structure of the KDP crystal with adjacent protons.

'neutral' boundary where the adjacent dipoles corresponding to these configurations are antiparallel to each other, and the 'polar' boundary with the parallel orientation of these dipoles. From the viewpoint of electrostatic energy, the 'polar' wall is more advantageous.

The scheme in Fig.3.6 shows good qualitative presentation on the structure of the domain boundary in the non-deuterated crystals of the $KH_2PO_4$ group only at low temperatures. The presence of the tunnelling effect of protons on hydrogen bonds in the non-deuterated crystals and also temperature different from zero will lead to disordering in the positions of protons on the hydrogen bonds. Evidently, this affects both the structure and surface density of energy of the domain wall. Besides, in real calculations in addition to the short-range interaction, which are taken into account with the help of the energy of the boundary configurations (Fig.3.5), it is also important to consider the electrostatic energy of interaction of the dipoles in the boundary.

Let us sequentially take into account the impact of the above factors on the parameters of the domain boundary in the $KH_2PO_4$ type crystals. Let us consider a flat domain wall in the infinite crystal with spontaneous polarization, oriented along the $z$ axis and normal to the wall coinciding with the $x$ axis. To describe the short-range interactions in the boundary, we use the Hamiltonian of the Izing model in the transverse field [6]:

$$H = -\Omega \sum_i X_i - \frac{1}{2} \sum_{ij} \mathcal{J}_{ij} Z_i Z_j. \tag{4.1}$$

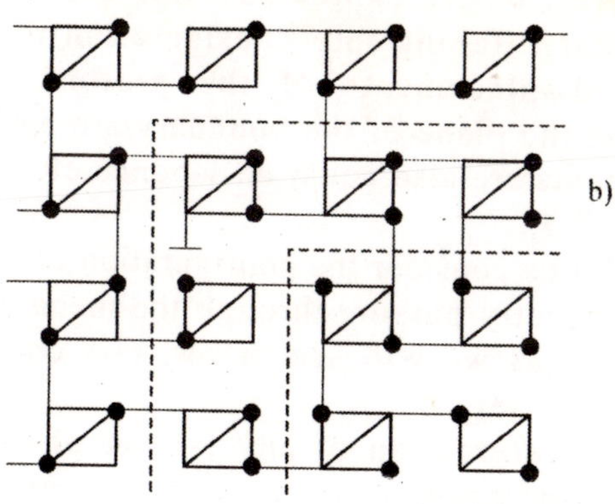


**Fig. 3.6.** The model of the domain boundary in the $KH_2PO_4$ crystal (indicated by the dashed line), consisting of Slater static configurations: (a) – the neutral boundary, (b)– the polar boundary [78].

Here $i$ is the number of the proton on the hydrogen bond, $\Omega$ is the tunnelling constant (integral), $\mathcal{J}_{ij}$ are the constants of quasi-spin interaction. The values of $\mathcal{J}_{ij}$ differ from zero only for the interaction of the nearest neighbours, and there are only two different interaction constants [6]. For the interaction of $x$–$y$, $y$–$x$ bonds $\mathcal{J}_{ij} = V$, and for the interaction of $x$–$x$, $y$–$y$ bonds $\mathcal{J}_{ij} = U$. $X_i$, $Z_i$ are quasi-spin operators describing the position of the proton on the hydrogen bond. The wave function of the proton on the hydrogen bond is modelled in the form of a linear combination of functions $\psi_i = a_i|\uparrow) + b_i|\downarrow)$, where $|\uparrow)$ and $|\downarrow)$ describe the position of the proton away from the boundary and are presented in the form of a linear combination of the functions $|\uparrow\rangle$ and $|\downarrow\rangle$ localized at 'upper' and 'lower' oxygen ions on the bond:

$$\left|\uparrow\right) = a_\infty\left|\uparrow\right\rangle + b_\infty\left|\downarrow\right\rangle, \quad \downarrow\right) = a_\infty\left|\downarrow\right\rangle + b_\infty\left|\uparrow\right\rangle, \\ \left(a_\infty^2 + b_\infty^2 = 1\right). \tag{4.2}$$

Coefficients $a_\infty$, $b_\infty$ describe the position of the proton on the bond away from the boundary, because taking tunnelling into account $|\langle Z\rangle_{\pm\infty}| \neq 1$.

The problem of finding the coordinate dependence $\langle Z_i\rangle$, describing the location of the proton on $i$-th hydrogen bond, is reduced to finding coefficients $a_i$ and $b_i$ since,

$$\langle Z_i\rangle = \left(a_i^2 - b_i^2\right)\left(a_\infty^2 - b_\infty^2\right). \tag{4.3}$$

Coefficients $a_\infty$, $b_\infty$ are expressed using the value of $\langle Z_i\rangle$ away from the boundary. Taking into consideration the symmetry it is assumed that displacements of the protons on the $x$-bonds equidistant from the plane of the boundary are equal in magnitude. Such displacements are also equal on $y$-bonds closely located to the same group of $PO_4$.

First of all, let us consider the configuration of the boundary with the plane of symmetry passing through the mean position of the $y$-bonds (Fig. 3.7) (as we will see below, this configuration is the main one in this case).

The effective interaction of any $y$-bond of the middle of the boundary layer (one chain of bonds) with adjacent $x$-bonds is equal to zero. Consequently, the displacement of the protons in the middle of the boundary layer depends only on the interaction of these $y$-bonds with each other and on the tunnelling effect of protons on hydrogen bonds. Assuming that the boundary is narrow, the

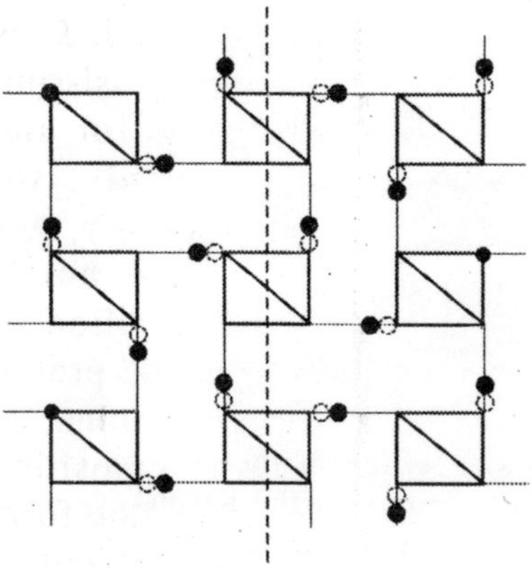

**Fig. 3.7.** The position of the protons in the domain boundary of the $KH_2PO_4$ crystal. Low temperatures. Basic configuration. Dashed circle – position of the protons according to Bjorkstam [78].

coefficients $a_i$ and $b_i$ can be calculated by the variational method. Let us introduce the variational parameters $a_1$ and $a_2$, $a_1$ for the bonds of the middle of the boundary layer, and $a_2$ for the rest. To facilitate numeration of the bonds, let us also introduce bands parallel to the plane of the boundary, with the width equal to half the size of the eiementary cell and with the boundaries of the bands passing through the middle of the oxygen octahedrons. Coefficients $a_i$ and $b_i$ are then assigned with the help of the following ratios:

a) the bonds of the middle of the boundary layer

$$\psi_i = a_1|\uparrow\rangle + \sqrt{1-a_1^2}|\downarrow\rangle, \psi_{i\pm1} = \sqrt{1-a_1^2}|\uparrow\rangle + a_1|\downarrow\rangle \tag{4.4}$$

b) other bonds

$$\begin{aligned} &\text{for } x\text{-bonds } m \geq 1 \quad b_m = a_2^m, \quad m \leq -1, \quad a_m = a_2^{|m|} \\ &\text{for } y\text{-bonds } m \geq 1 \quad b_m = a_2^m, \quad m \leq -1, \quad a_m = a_2^{|m|-1}, \end{aligned} \tag{4.5}$$

where $m$ is the number of the band. Taking into account (4.5), the energy directly linked with the position of the protons in the boundary layer, $H_1 = \langle \psi_i^* | H_i | \psi_i \rangle \int \langle \psi_i^* | \psi_i \rangle$ has the form of

$$\begin{aligned} H_1 = &\frac{K_2 - 16\Omega a_2 + Q[48V + 16U]a_2^2 - 16\Omega a_2^2 + 8\Omega a_2^3}{1 + 16a_2R} + \\ &+ K_1 - 4\Omega a_1\sqrt{1-a_1^2} + 2U\left(2a_1^2 - 1\right)^2. \end{aligned} \tag{4.6}$$

Here $K_1 = (4V + 2U)\cdot Q$, $K_2 = (4V + 2U)\cdot Q$, $Q = \left(a_\infty^2 - b_\infty^2\right)$, $R = 2a_\infty b_\infty$.

Similarly, the following ratios can be introduced for the configuration of the boundary with the plane of symmetry passing through the middle of the $x$-bonds (Fig.3.8):

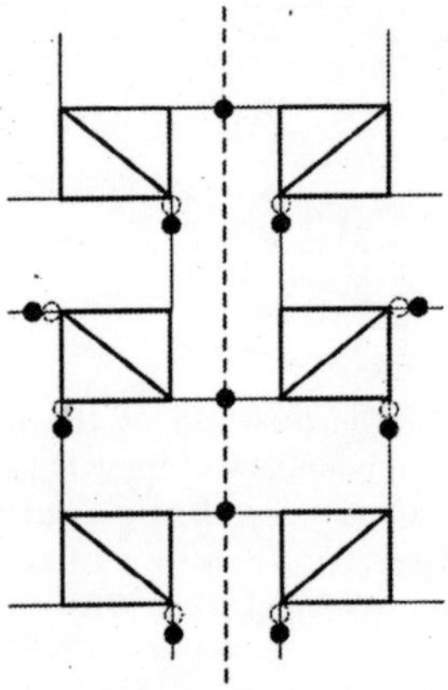

**Fig. 3.8.** Position of the protons in the domain boundary of the $KH_2PO_4$ crystal. Lower temperatures, saddle configuration.

a) the bonds of the middle of the boundary layer (two chains of $y$-bonds, connected by $x$-bonds):

for the bonds of the left chain

$$\psi_i = \tilde{a}_1 |\uparrow\rangle + \sqrt{1-\tilde{a}_1^2}\,|\downarrow\rangle, \quad \psi_{i\pm1} = \sqrt{1-\tilde{a}_1^2}\,|\uparrow\rangle + \tilde{a}_1 |\downarrow\rangle,$$

for the bonds of the right chain

$$\psi_i = \sqrt{1-\tilde{a}_1^2}\,|\uparrow\rangle + \tilde{a}_1 |\uparrow\rangle, \quad \psi_{i\pm1} = \tilde{a}_1 |\uparrow\rangle + \sqrt{1-\tilde{a}_1^2 |\uparrow\rangle},$$

for the connecting bonds

$$\psi_i = \frac{1}{\sqrt{2}}|\uparrow\rangle + \frac{1}{\sqrt{2}}|\downarrow\rangle; \tag{4.7}$$

b) other bonds

$$\begin{aligned} &\text{for } x\text{-bonds } m>1, \quad b_m = \tilde{a}_2^{m-1}, \quad m \le -1, \quad a_m = \tilde{a}_2^{|m|}, \\ &\text{for } y\text{-bonds } m>1, \quad b_m = \tilde{a}_2^{m-1}, \quad m < -1, \quad a_m = \tilde{a}_2^{|m|-1}. \end{aligned} \tag{4.8}$$

The energy

$$\begin{aligned} \tilde{H}_1 = {} & \frac{\tilde{K}_2 - 16\Omega\tilde{a}_2 + Q[48V + 24U]\tilde{a}_2^2 - 16\Omega\tilde{a}_2^2 + 8\Omega\tilde{a}_2^3}{1+16a_2R} + \\ & + \tilde{K}_1 - 8\Omega\tilde{a}_1\sqrt{1-\tilde{a}_1^2} + 4U\left(2\tilde{a}_1^2 - 1\right)^2, \end{aligned} \tag{4.9}$$

where $\tilde{K}_1 = 4\Omega R + (12V + 16U)Q$, $\tilde{K}_2 = (4V + 2U)Q$.

When taking into account the energy of the dipole–dipole interaction, the dipole moment of the complex K–$PO_4$ will be assumed as a point one and located in the centre of the oxygen octahedron for simplicity of considerations. Four nonequivalent complexes K–$PO_4$ are numerated by the indices $\mu,\nu = 1, 2, 3, 4$. Their position with respect to each other is determined by the following matrices

$$s_{1\mu\nu} = \frac{x_{\mu\nu}}{a_x} = \begin{pmatrix} 0 & 1/2 & 0 & 1/2 \\ & 0 & -1/2 & 0 \\ & & 0 & 1/2 \\ & & & 0 \end{pmatrix},$$

$$s_{2\mu\nu} = \frac{y_{\mu\nu}}{a_y} = \begin{pmatrix} 0 & 1/2 & 1/2 & 0 \\ & 0 & 0 & -1/2 \\ & & 0 & -1/2 \\ & & & 0 \end{pmatrix}, \qquad (4.10)$$

$$s_{3\mu\nu} = \frac{z_{\mu\nu}}{a_z} = \begin{pmatrix} 0 & 1/2 & 1/4 & -1/4 \\ & 0 & -1/4 & -3/4 \\ & & 0 & -1/2 \\ & & & 0 \end{pmatrix}.$$

To calculate the energy of dipole–dipole interaction, it is convenient to find in advance electric fields generated by individual dipole planes, parallel to the plane of the domain boundary in the location of an arbitrary dipole.

The latter can be calculated using the procedure described in section 3.2 and utilized for calculation of the fields in the crystal of barium titanate. Due to high tetragonality of the elementary cell ($a_1/a_3= a_2/a_3= 1.07$) it is not possible here to calculate the lattice factor immediately for the entire plane. In this case, the factor is found by direct summation using equations (2.16) and (2.17), which yields the following values for the given structure: $I(0)=5.357$, $I(1)=I(-1)=1.947$, $I(2)=I(-2)=-0.105$. And at $I(3)=I(-3)$ the values are negligible.

Let us assume that the interaction of the proton subsystem with the dipole complexes is determined by the rigid local bond:

$$P=\frac{P_0}{4}\sum_{i=1}^{4}\langle Z_i\rangle, \qquad (4.11)$$

where $P_0$ is the dipole moment of the complex away from the boundary, and $\langle Z_i\rangle$ are the mean displacements of the protons on the hydrogen bonds linked to the given complex. Then, in accordance with (4.7), (4.8) the value of the dipole moment of a single complex K–$PO_4$ in the $m$-th layer of the main configuration of the wall is

$$P(m)=\frac{P}{4}\text{sign}(m)\cdot\left[3\left(1-2a_2^{2|m|}+\left(1-2a_2^{2|m|+2}\right)\right)\right], m\neq 0. \qquad (4.12)$$

Taking into account (4.12) and specific values of the lattice factor, the surface density of the dipole energy of the main configuration of the wall is

$$H_2=\frac{4P^2}{a^5}\left\{A+Ba_2^2+Ca_2^4\right\}, \qquad (4.13)$$

where $A$ = 6.14, $B$ = 20.145, $C$ = 19.75.

Similarly, the dipole moment of the complex of the heavy ions of the $m$-th layer for the saddle configuration is

$$P(m)=\frac{P}{4}\text{sign}(m)\cdot\left[3\left(1-2\tilde{a}_2^{2|m|-2}\right)+\left(1-2a_2^{2|m|}\right)\right],$$
$$|m|>1, P(-1)=P(1)=\frac{P}{4}\cdot\left[\left(1-2\tilde{a}_2^2\right)\right], \qquad (4.14)$$

from which the surface density of the dipole energy for the saddle configuration of the wall is

$$\tilde{H}_2=\frac{4P^2}{a^5}\left\{\tilde{A}+\tilde{B}\tilde{a}_2^2+\tilde{C}\tilde{a}_2^4\right\}, \qquad (4.15)$$

where $\tilde{A}$ = 9.6, $\tilde{B}$ = 25.9, $\tilde{C}$=16.4.

Summation of $H_1/a^2$ and $H_2$, $\tilde{H}_1/a^2$ and $\tilde{H}_2$ gives the total energy of the main and saddle configurations of the domain wall respectively. In this case, the form of the transition layer is determined by the variation in respect of the introduced parameter.

The use of the energy values of the boundary configuration of type 2 $\varepsilon_0^H$ = 64 K, and Takagi's defect $W_H$ = 680 K, $\Omega_H$ = 86 K [80–85], which enables to find the energy constants (4.6) and (4.9), gives the following values of variation parameters determining the structure of the wall: $a_1$=0.13, $a_2$=0.20, $Q$=0.98, $\tilde{a}_1$ =0.13, $\tilde{a}_2$ =0.28. In this case, the surface density of the energy of the wall for the main and saddle configuration of the wall are equal to $\gamma_0$=25 erg·cm$^{-2}$, $\gamma$=53 erg·cm$^{-2}$ [79].

### 3.5. Temperature dependence of the lattice barrier in crystals of the $KH_2PO_4$ group

The consideration of the structure of boundary configurations made above with a relatively detailed investigation of the proton position at each of the boundary bonds is evidently applicable only for the case of relatively low temperatures. With the increase of temperature and, therefore, of the number of boundary bonds, it becomes more and more difficult to follow the details of displacement of the growing number of the particles. In this case, it is more realistic to use the approach based on the consideration of the mean characteristics. Such an approach can be represented, for example, by the use of the approximation of the mean (molecular) field in the already discussed Hamiltonian (4.1).

Utilizing the symmetry of the problem, i.e. homogeneity in a plane parallel to the plane of the domain boundary, and carrying out averaging in respect of such planes, in the approximation of the molecular field the Hamiltonian (4.1) can be written in the form of the sum of Hamiltonians

$$\begin{aligned} H_{i,n} &= \frac{1}{2}\left[\mathcal{J}\langle Z_n^2\rangle + A\left(\langle Z_{n-1}\rangle + \langle Z_{n+1}\rangle\right)\langle Z_n\rangle\right] - \\ &-\Omega X_{i,n} - \left[\mathcal{J}\langle Z_n\rangle + A\left(\langle Z_{n-1}\rangle + \langle Z_{n+1}\rangle\right)\right]\cdot Z_{i,n}. \end{aligned} \tag{5.1}$$

Here $Z_{i,n}$, $X_{i,n}$, are the operators of quasi-spin of the $i$-th bond belonging to the $n$-th plane, parallel to the domain boundary. The mean value of the quasi-spin $\langle Z_n\rangle$ depends on the number $n$ of the plane (layer) and determines the degree of ordering (polarization) in the given location of the crystal.

When writing (5.1) it is assumed that the dependence of the constant of quasi-spin interaction $\mathcal{J}_{ij}$ on the numbers of interacting quasi-spins is reduced to the dependence of the constant on the direction of interaction. In (5.1) constant $\mathcal{J}$ is the cumulative constant of interaction of the quasi-spin with neighbours in the direction parallel to the plane of the boundary, and constant $A$ – with neighbours in the direction perpendicular to the plane of the boundary.

It was shown in the previous section that the electric fields, generated by different dipole planes, in the approximation of the rigid bond of the proton subsystem and the system of heavy ions, combining equation (2.2) and (4.11), can be written in the form of the product $\sim I(n-m)\ \langle Z_n\rangle$, where $I(n-m)$ is the corresponding structural factor. At same time the energy of interaction of the given dipole with all dipoles of the $m$-th plane is $\sim I(n-m)\ \langle Z_n\rangle\langle Z_m\rangle$, i.e. it has the same structure as the short-range part of the interaction of quasi-spins. Taking into account the short-range nature of the electric field of the dipole plane, its exponential decrease with distance (see (2.12) and calculations, for example in the previous section) in the general expression for the energy of dipole-dipole interaction, we can retain only the terms with $m=n$ and $m=n-1$, $n+1$. As a result as well as in the case of the short-range interaction, formally we have the interaction with the nearest neighbour although in fact it represents the interaction with all the dipoles of the given plane. The resultant identity of the structure of the local dipole–dipole interaction for the system with the uniform distribution of polarization in the plane with the normal to the vector $\mathbf{P}_0$ allows to add them up and to consider constants $\mathcal{J}$ and $A$ as the cumulative constants that take into account both local short-range and dipole–dipole interactions.

In exact calculations of the parameters of the boundary at $T \neq 0$ on the basis of (5.1), depending on the situation, it is necessary to calculate either the thermodynamic potential or free energy. Calculation of the corresponding statistical sum assuming that the system under consideration is investigated at a constant pressure, leads to the following expression for the surface density of the energy of the boundary [86.86]:

$$\gamma = \frac{1}{S}\sum_n \left\{ \frac{\mathcal{J}}{2}[Z_n S_n - Z_\infty \cdot S_\infty] - -T\left[ \ln 2\operatorname{ch}\left( \frac{\mathcal{J}\sqrt{q^2 + S_n^2}}{T} \right) - \ln 2\operatorname{ch}\left( \frac{\mathcal{J}\sqrt{q^2 + S_\infty^2}}{T} \right) \right] \right\}. \tag{5.2}$$

The following notations were used when writing (5.2)

$$q = \Omega / \mathcal{J}, \quad S_n = Z_n + \frac{A}{\mathcal{J}}(Z_{n+1} + Z_{n-1}),$$
$$S_\infty = \left(1 + \frac{2A}{\mathcal{J}}\right) \cdot Z_\infty, \quad Z_n \equiv \langle Z_n \rangle, \tag{5.3}$$

where $Z_\infty$ is the mean value of the quasi-spin away from the boundary, and $S$ is the area of the side surface of the elementary cell parallel to the plane of the domain wall and falling onto a single quasi-spin chain.

Self-congruent values $Z_n$ and $Z_\infty$ are determined in the general case from the minimality conditions $\partial\gamma/\partial Z_n = 0$ and $\partial\gamma/\partial Z_\infty = 0$ and comply with the following respective equations

$$Z_n\sqrt{q^2+S_n^2}=S_n\cdot\text{th}\left(\frac{\mathcal{J}\sqrt{q^2+S_n^2}}{T}\right),$$
$$Z_\infty\cdot\sqrt{q^2+S_\infty^2}=S_\infty\cdot\text{th}\left(\frac{\mathcal{J}\sqrt{q^2+S_\infty^2}}{T}\right). \tag{5.4}$$

In direct calculations of the structure of boundary configurations, as in the case of barium titanate (see section 3.3), the first of the equations (5.4) can be used for the middle of the boundary layer with $n$=0, ±1. For the remaining part of the boundary, the dependence of $Z_n$ on $|n|$ can be simulated by the expression

$$Z_n=Z_\infty\cdot\left[1-A\cdot\exp\left(-|n|\cdot\lambda\right)\right], \tag{5.5}$$

where parameter $\lambda$ is determined from the self-congruent condition, based on the application of distribution (5.5) in the general equation (5.4) for high $n$, which leads to the following equation for determination of $\lambda$:

$$\sqrt{q^2+S_\infty^2}=\left(\text{th}\frac{\mathcal{J}}{T}\sqrt{q^2+S_\infty^2}-\frac{S_\infty\cdot Z_\infty}{\sqrt{q^2+S_\infty^2}}+\right.$$
$$\left.+\frac{\mathcal{J}\cdot S_\infty^2/T}{\sqrt{q^2+S_\infty^2}}\cdot\text{ch}^2\frac{\mathcal{J}\sqrt{q^2+S_\infty^2}}{T}\right)\cdot\left(1+2\frac{A}{\mathcal{J}}\text{ch}\,\lambda\right). \tag{5.6}$$

Coefficient $A$ in (5.5) is determined from the condition of joining of solutions (5.4) and (5.5) at $n$ = 1.

Using the calculated value $\Omega_H$ = 72 K, $\Omega_D$ = 0, one can find the total value of the constants ($\mathcal{J}$+2$A$) from the condition for the transition temperature $\frac{\Omega}{(\mathcal{J}+2A)}=\text{th}(\Omega/T_c)$, resulting from (5.4) at $Z_n=Z_{n+1}=Z_{n-1}=Z_\infty$. This gives $\mathcal{J}+2A_H$=139 K and $(\mathcal{J}_D+2A_D)=T_c^D$=213 K. Assuming that the short-range local interaction is symmetric and the contribution of the dipole–dipole interaction to constants $\mathcal{J}$ and $A$ is determined by the ratio of the factors $I$(0) and $I$(±1), we can also find the value of the individual constants, which turn out to be as follows: $\mathcal{J}_H$=113 K, $A_H$=13 K, $\mathcal{J}_D$=173 K, $A_D$=20 K.

Numerical calculations of the structure and surface density of the energy of the boundary configurations $a$ and $b$ (Fig.3.1) denoted below as II and I respectively using the obtained values of $\mathcal{J}$, $A$, $W$ on the basis of the ratios (5.4)–(5.6) show the following (Fig. 3.9, 3.10) [87,88].

Up to the immediate vicinity of $T_c$($\sim$2÷3 K) for crystals of $KH_2PO_4$ and $KD_2PO_4$ $\lambda > 2$, $\alpha_H^I = 1.15$, $\alpha_H^{II} = 2.45$, $\alpha_D^I = 1.0$, $\alpha_D^{II} = 2.17$. Thus, for almost all $n \geq 2$ for the both types of the boundary configurations here $Z_n \simeq Z_\infty$ and, consequently, in the

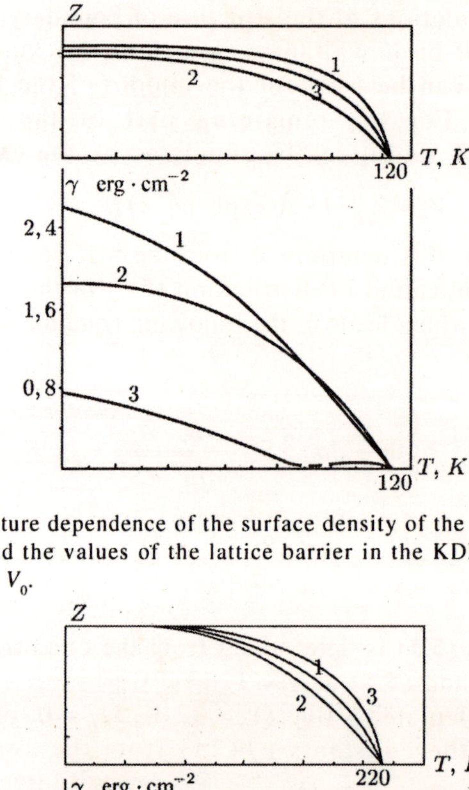


**Fig. 3.9.** Temperature dependence of the surface density of the energy of boundary configurations and the values of the lattice barrier in the KDP crystal. 1) $Z_{1I'}$, $\gamma_{I'}$, 2) $Z_{1II'}$, $\gamma_{II'}$, 3) $Z_\infty$, $V_0$.

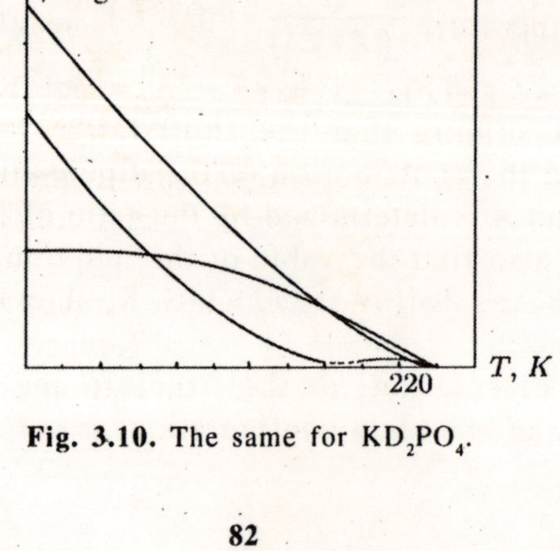


**Fig. 3.10.** The same for $KD_2PO_4$.

entire mentioned temperature range the domain boundaries in these crystals remain narrow.

A characteristic feature of Figs. 3.9 and 3.10 is the intersection of the dependences $\gamma_I$ and $\gamma_{II}$, i.e. the change of the type of the main configuration of the domain wall at some temperature $T_0$. At $T > T_0$ the main configuration is the configuration of type I, at $T < T_0$ it is the configuration of type II.

To find out all possible reasons for alternation in the type of structure of the boundary with the temperature change, let us compare the energy of the configuration of the type I and II for narrow boundaries (Fig.3.11), whose width is comparable with the lattice constant and permits simple analytical estimates. To simplify considerations, let us make estimates using the continuous model.

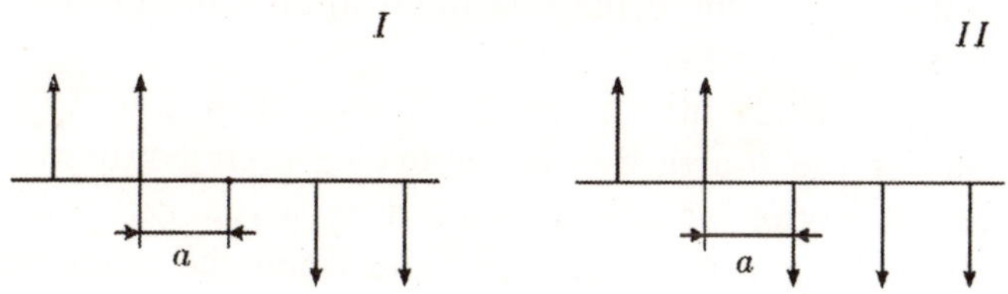


**Fig. 3.11.** Structure of the narrowest configurations of the domain wall with extreme energy values.

The volume contribution to $\gamma$ for configuration I is the quantity $\alpha P_0^2 a$, whereas the correlation term is equal to $\frac{\varkappa}{4}\frac{P_0^2}{a}$. Their comparison taking into account the expression for the half width of the wall $\delta = \sqrt{2\varkappa/\alpha} = a$ shows that the volume contribution prevails here (fourfold). In the case of configuration II the volumetric contribution is equal to zero, whereas the correlation contribution is equal to $\frac{\varkappa}{2}\frac{P_0^2}{a}$. We can see that the temperature dependence of the quantities $\gamma_I$ and $\gamma_{II}$ differs: $\gamma_I \sim \Delta T^2$, $\gamma_{II} \sim \Delta T$ ($\Delta T = T_c - T$). At some temperature $T_0$, where $\alpha P_0^2 a = \frac{\varkappa P_0^2}{2a}$, they intersect and this is the temperature of structural rearrangement in the domain boundary: configuration of type I exists above $T_0$, and configuration of type II exists below $T_0$.

As it can be seen the mentioned above structural rearrangement in the boundary can be explained by the differences in the temperature dependences of the volumetric and correlation contributions to the surface density of the boundary energy. It should be noted that the possible change of the type of the main configuration for the same reason it is pointed out, in particular, in the Frenkel–Kontorova model [73].

### 3.6 Influence of tunnelling on the structure of domain boundaries in ferroelectrics of the order–disorder type

Comparison of the results of calculation of parameters of domain walls for $KH_2PO_4$ and $KD_2PO_4$ crystals depicted in Figs. 3.9 and 3.10, shows the influence of tunnelling of the protons on the hydrogen bonds on the structure and surface density of the boundary energy [79, 89–91]. To detect this effect in a more obvious form it is convenient to consider it in the area where it permits analytical description. The relatively close vicinity of $T_c$ where the continual approximation can be used is such an area in our case.

Expansion of (5.2) into a series in respect of low $Z_n$ and $Z_\infty$ up to the terms of the fourth degree inclusive taking into account the difference analogue of the second derivative $(Z_{n+1} - 2Z_n + Z_{n-1})/a^2 \to d^2Z/dx^2$, where $a$ is the distance between the adjacent planes, after transition to the continual limit enables $\gamma$ to be presented in the form of

$$\gamma = \frac{1}{S}\int\left\{\frac{\alpha}{2}\left(Z^2 - Z_\infty^2\right) + \frac{\beta}{4}\left(Z^4 - Z_\infty^4\right) + \frac{\varkappa}{2}\left(\frac{dZ}{dx}\right)^2\right\}\frac{dx}{a}, \tag{6.1}$$

where

$$\alpha = (\mathcal{J} + 2A) - \frac{(\mathcal{J} + 2A)^2}{\Omega}\cdot \mathrm{th}\frac{\Omega}{T}, \tag{6.2}$$

$$\beta = \frac{(\mathcal{J} + 2A)^4}{2T_c \cdot \Omega}\left[\frac{T_c}{\Omega}\mathrm{th}\frac{\Omega}{T_c} - \frac{1}{\mathrm{ch}^2\Omega/T_c}\right], \tag{6.3}$$

$$\varkappa = A\cdot a^2\left[2\frac{(\mathcal{J} + 2A)}{\Omega}\mathrm{th}\frac{\Omega}{T_c} - 1\right] \equiv a^2 \cdot A. \tag{6.4}$$

When writing (6.2)–(6.4) it is assumed that only coefficient $\alpha$ depends on temperature in an explicit manner, and the coefficients $\alpha$, $\beta$, $\varkappa$ are normalized in the corresponding manner, since, for example, $\alpha/Sa$ has the dimensionality of the volume density of energy.

In the vicinity of $T_c$

$$\alpha \simeq \alpha_0 (T - T_0), \quad \alpha_0 = \frac{(\mathcal{J} + 2A)^2}{\mathrm{ch}^2 \Omega / T_c} \cdot \frac{1}{T_c^2}, \tag{6.5}$$

where the value of $T_c$ itself is determined by the conventional ratio $\Omega/(\mathcal{J}+2A) = \mathrm{th}\Omega/T_c$. In this case, the structure and half width of the domain wall have the form [89]:

$$Z(x) = Z_\infty \cdot \mathrm{th}\frac{x}{\delta}, \; \delta = a\sqrt{\frac{2A}{T - T_c} \cdot \frac{T_c \cdot \mathrm{ch}\Omega / T_c}{(\mathcal{J} + 2A)}}. \tag{6.6}$$

At low $\Omega$

$$\delta \simeq a\sqrt{\frac{2A}{T - T_c}} + a\sqrt{\frac{2A}{T - T_c}} \cdot \frac{\Omega^2}{2(\mathcal{J} + 2A)^2}. \tag{6.7}$$

The surface density of the energy of the domain wall in compliance with (6.1) and (6.2)–(6.4) is

$$\begin{aligned} \gamma &= \frac{2}{3}\alpha Z^2 \delta \cdot a^{-3} = \frac{2\sqrt{2}}{3}\alpha^{3/2}\varkappa^{1/2}\beta^{-1}a^{-3} = \\ &= \frac{4\sqrt{2}}{3} \cdot \frac{A^{1/2}(T - T_c)^{3/2}\Omega^2}{a^2(\mathcal{J} + 2A) \cdot T_c^2 \cdot \mathrm{ch}^3 \Omega / T_c} \times \\ &\qquad \times \left[\frac{T_c}{\Omega}\mathrm{th}\frac{\Omega}{T_c} - \frac{1}{\mathrm{ch}^2 \Omega / T_c}\right]^{-1}. \end{aligned} \tag{6.8}$$

At low $\Omega$

$$\gamma = \sqrt{2} \cdot \Delta T^{3/2} \cdot A^{1/2} (\mathcal{J} + 2A)^{-1} / a^2 \cdot ch^3 \frac{\Omega}{(\mathcal{J} + 2A)}, \qquad \Delta T_c = T - T_c. \tag{6.9}$$

In this case, the dimensionless order parameter in the volume of the domain is

$$Z = \sqrt{\alpha / \beta} = \frac{\sqrt{3} \cdot T_c^{1/2} (T - T_c)^{1/2} \left[1 - \frac{\Omega^2}{T_c^2}\right]^{1/2}}{\mathcal{J} + 2A}, \tag{6.10}$$

and, consequently, the derivative characterizing the curvature of the profile of the domain wall is:

$$\frac{dZ}{dx} \simeq \frac{Z}{\delta} = \frac{\alpha}{\sqrt{2\varkappa\beta}} = \frac{(T - T_c)}{\sqrt{A \cdot T_c \cdot a}} \cdot \left[1 - \frac{\Omega^2}{T_c^2}\right]. \tag{6.11}$$

As it can be seen from Figs. 3.12 and 3.13, tunnelling that differs from zero increases the width of the domain wall and reduces the density of its surface energy.

As numerical estimates show at $\Delta T \sim 10$ K and $a \sim 10^{-7}$ cm, the width of the domain wall in DKDP is $\delta_D \simeq 2 \cdot 10^{-7}$ cm, at the same time the surface density of its energy is $\gamma_D \simeq 6 \cdot 10^{-2}$ erg·cm$^{-2}$. In KDP crystal at the same distance from $T_c$ $\delta_H = 2.5 \cdot 10^{-7}$ cm, $\gamma_H \simeq 4 \cdot 10^{-2}$ erg·cm$^{-2}$, which is in good agreement with the results of numerical calculations of the previous section. The mentioned agreement is conditioned by the possibility of using here the continual approximation, the transition to which gives the relative error of $\Delta\gamma/\gamma \simeq a^2/2\delta^2 << 1$, as indicated in particular by estimates for $\gamma$.

As we will see in chapter 5, a not too large increase of the width of the domain wall at $\Omega \neq 0$ can result in an extremely large increase of its mobility.

The link of the parameters of continual approximation to the microscopic model, found in this section, in particular, in the

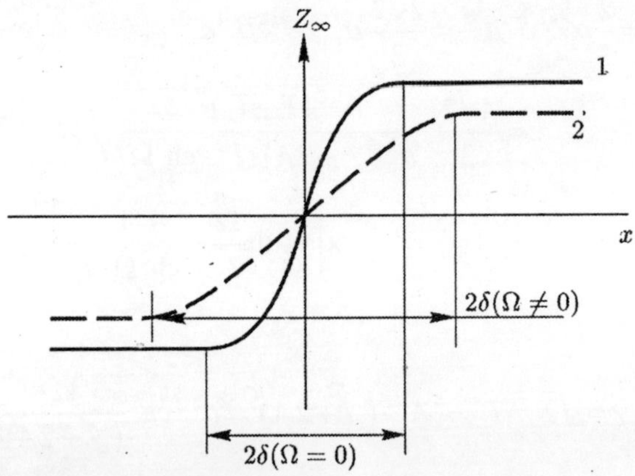


**Fig. 3.12.** Alternation of the order parameter in the boundary for different values of the tunnelling constant. 1 — $\Omega = 0$, 2 — $\Omega \neq 0$.

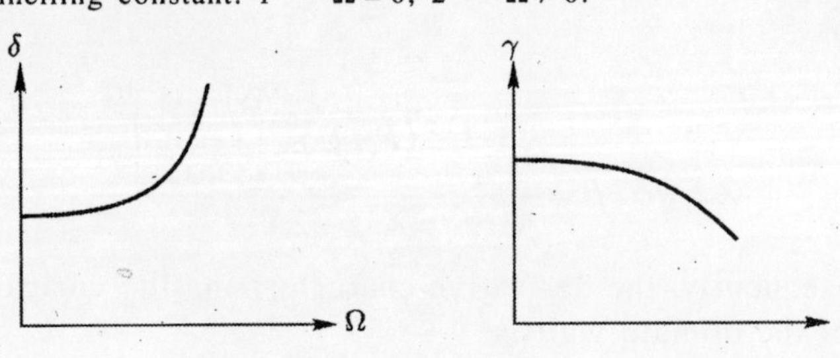


*a)* *b)*

**Fig. 3.13.** Change of the width of the domain wall (a) and surface density of its energy (b) in relation to the value of the tunnelling integral.

expression for the width of the wall (6.7) makes it possible to supplement the interpretation of structural rearrangement in the boundary at $T = T_0$, described in the previous section. Indeed, from the equality condition $\gamma_I = \gamma_{II}$ at $T = T_0$ it follows that the structural rearrangement in the boundary takes place at $\delta \simeq a$. To simplify considerations, assuming in the expression for $\delta$ (6.7) that $\Omega = 0$, we see that at the first approximation the rearrangement in the boundary should take place at the temperature $T_0 = T_c - 2A = \mathcal{J}$. But this is just the temperature of ordering in the layer parallel to the boundary. Since due to symmetry the interaction of the central layer in configuration I with the neighbours is always equal to zero, then in accordance with the estimates the structural rearrangement in the boundary considered here may be understood as ordering as a result of the phase transformation in the central layer of the boundary.

### 3.7. Structure of the domain boundaries in $KH_{2x}D_{2(1-x)}PO_4$ solid solutions

Discussion identical to that in the previous section can be also carried out for $KH_2PO_4$–$KD_2PO_4$ solid solutions. The surface density of energy of the domain boundary in this case, written for simplicity in the approximation of $\Omega = 0$, has the form

$$\gamma = \frac{1}{S}\sum_n \left\{ x \cdot \frac{\mathcal{J}_H}{2} Z_n \cdot S_n^H + (1-x) \cdot \frac{\mathcal{J}_D}{2} Z_n \cdot S_n^D - \right.$$

$$\left. -T \cdot x \ln 2\mathrm{ch} \frac{\mathcal{J}_H S_n^H}{T} - T \cdot (1-x) \ln 2\mathrm{ch} \frac{\mathcal{J}_D S_n^D}{T} \right\} - \gamma(Z_\infty). \tag{7.1}$$

Here $\gamma(Z_\infty)$ is the density of the free energy of the homogeneous state implemented at $n \to \infty$, i.e. away from the boundary. Quantity $x$ determines the degree of deuteration of the crystal, $\mathcal{J}_{H,D}$ and $A_{H,D}$ are the effective constants of interaction of the quasi-spin with neighbours in the direction parallel to the plane of the boundary and normal to it for the case in which the given cell contains the atoms of hydrogen and deuterium respectively

$$S_n^{H,D} = Z_n + \frac{A_{H,D}}{\mathcal{J}_{H,D}} \cdot (Z_{n+1} + Z_{n-1}) \tag{7.2}$$

The equilibrium equation, determining the coordinate dependence of the quasi-spin in the boundary, is obtained from (7.1) by its minimization in respect of $Z_n$ and has the following form

$$x\cdot \mathcal{J}_H\cdot S_n^H+(1-x)\cdot \mathcal{J}_D\cdot S_n^D-x\mathcal{J}_H \operatorname{th}\left(\frac{\mathcal{J}_H S_n^H}{T}\right)-$$
$$-xA_H \operatorname{th}\left(\frac{\mathcal{J}_H S_{n+1}^H}{T}\right)-xA_H \operatorname{th}\left(\frac{\mathcal{J}_H S_{n-1}^H}{T}\right)-$$
$$-(1-x)\mathcal{J}_D \operatorname{th}\left(\frac{\mathcal{J}_D S_n^D}{T}\right)-(1-x)A_D \operatorname{th}\left(\frac{\mathcal{J}_D S_{n+1}^D}{T}\right)- \tag{7.3}$$
$$-(1-x)A_D \operatorname{th}\left(\frac{\mathcal{J}_D S_{n-1}^D}{T}\right)=0$$

In the homogeneous case equation (7.3) changes to

$$x\cdot I_H\cdot Z_\infty+(1-x)I_D\cdot Z_\infty=$$
$$=I_H\cdot x\cdot \operatorname{th}\frac{I_H\cdot Z_\infty}{T}+I_D\cdot(1-x)\cdot \operatorname{th}\frac{I_D\cdot Z_\infty}{T}, \tag{7.4}$$

from which at low $Z_\infty$, we obtain an expression for determination of the temperature of the phase transition of partially deuterated crystal [92,93].

$$T_c=\frac{(1-x)\cdot I_D^2+x\cdot I_H^2}{(1-x)\cdot I_D+x\cdot I_H} \tag{7.5}$$

where $I_{H,D}=\mathcal{J}_{H,D}+2A_{H,D}$.

Calculations of the structure and density of the surface energy of the boundary in the general case are relatively cumbersome. In some cases, however, it is sufficient to describe the boundary within the framework of the already continual approximation. The thickness of the boundary in this case, for example, for a ferroelectric with the phase transition of second order, is $\delta=\sqrt{2\varkappa/\alpha}$. To determine the correlation constant $\varkappa$ and the temperature-dependent coefficient of expansion of free energy $\alpha$, let us express them, as in section 3.5, with the help of the available parameters $\mathcal{J}_{H,D}$ and $A_{H,D}$. For this purpose, let us go over to a continual limit in (7.1). Introducing here the difference combinations $\Delta Z_n=Z_{n+1}-2Z_n+Z_{n-1}$ into the terms with coefficients $A_{H,D}$, and taking into account that the quantity $\Delta Z_n/a^2$ is the difference analogue of the second derivative $d^2Z/dx^2\equiv Z''$, after going over to the approximation of the continuous medium, where $Z = Z(x)$, we have the following expression:

$$\gamma = \frac{1}{S}\int \frac{dx}{a}\left\{ x\left[\frac{I_H}{2}Z^2 + \frac{A_H}{2}a^2 Z\cdot Z''\right] + \right.$$
$$+(1-x)\cdot\left[\frac{I_D}{2}Z^2 + \frac{A_D}{2}a^2 Z\cdot Z''\right] -$$
$$-T\cdot x \ln 2\mathrm{ch}\left(\frac{I_H\cdot Z + A_H\cdot a^2\cdot Z''}{T}\right) - \tag{7.6}$$
$$\left. -T\cdot(1-x)\ln 2\mathrm{ch}\left(\frac{I_D\cdot Z + A_D\cdot a^2\cdot Z''}{T}\right)\right\} - \gamma(Z_\infty).$$

After expansion of logarithmic terms in (7.6) into a series

$$\gamma = \frac{1}{S}\int \frac{dx}{a}\left\{ x\left[\frac{I_H}{2}Z^2 + \frac{A_H}{2}a^2 Z\cdot Z''\right] + \right.$$
$$+(1-x)\cdot\left[\frac{I_D}{2}Z^2 + \frac{A_D^2}{2}a^2 Z\cdot Z''\right] -$$
$$-x\left[\frac{I_H^2}{2T}\cdot Z^2 - \frac{I_H\cdot A_H\cdot a^2}{T}\cdot Z''\right] - \tag{7.7}$$
$$\left. -(1-x)\left[\frac{I_D^2}{2T}\cdot Z^2 - \frac{I_D\cdot A_D\cdot a^2}{T}\cdot Z''\right]\right\}.$$

Taking into account integration by parts $\int_{-\infty}^{\infty} ZZ''\partial x = \int_{-\infty}^{\infty}(dZ/dx)^2\,dx$, after collecting together the terms with $Z^2$ and $(dZ/dx)^2$, the expression for $\gamma$ is written in the form:

$$\gamma = \frac{1}{S}\int \frac{dx}{a}\left\{\left[x\cdot I_H + (1-x)\cdot I_D - x\frac{I_H^2}{T} - (1-x)\frac{I_D^2}{T}\right]\frac{Z^2}{2} + \right.$$
$$+\frac{1}{2}\left[-xA_H\cdot a^2 - (1-x)A_D\cdot a^2 + x\frac{2A_H I_H a^2}{T} + \right. \tag{7.8}$$
$$\left.\left. +(1-x)2\frac{A_D I_D a^2}{T}\right]\left(\frac{dZ}{dx}\right)\right\} - \gamma(Z_\infty).$$

Comparing (7.8) with the continual expression for $\gamma$, written in terms of coefficients $\alpha$ and $\varkappa$, where

$$\gamma = \frac{1}{S}\int\left\{\frac{\alpha}{2}\left(Z^2 - Z_\infty^2\right) + \frac{\varkappa}{2}\left(\frac{dZ}{dx}\right)^2\right\}\frac{dx}{a}, \tag{7.9}$$

we can see that here

$$\alpha = x \cdot I_H + (1-x)\cdot I_D - x\frac{I_H^2}{T} - (1-x)\frac{I_D^2}{T} \tag{7.10}$$

or, taking into account the expression for $T_c$(7.5)

$$\alpha = x\cdot\frac{I_H^2}{T_c} + (1-x)\cdot\frac{I_D^2}{T_c} - x\frac{I_H^2}{T} - (1-x)\frac{I_D^2}{T} = \alpha_0(T - T_c), \tag{7.11}$$

$$\alpha_0 = \frac{x\cdot I_H^2 + (1-x)I_D^2}{T_c^2}, \tag{7.12}$$

$$\frac{\varkappa^2}{a} = \frac{2x\cdot I_H\cdot A_H}{T_c} + \frac{2(1-x)I_D A_D}{T_c} - x\cdot A_H - (1-x)A_D. \tag{7.13}$$

Hence the width of the boundary [94] is:

$$\delta = a\left\{\frac{4xI_H A_H + 4(1-x)I_D A_D - 2A_H T_c - 2(1-x)A_D T_c}{\left(xI_H^2 + (1-x)I_D^2\right)(T - T_c)}\cdot T_c\right\}^{1/2}. \tag{7.14}$$

At $x = 1$ ($I_H \equiv I = T_c$, $A_H \equiv A$) the expression (7.14) changes to the previously derived expression (5.6), written for $\Omega = 0$.

Chapter 4

# Interaction of domain boundaries with crystalline lattice defects

## 4.1 INTERACTION OF A FERROELECTRIC DOMAIN BOUNDARY WITH A POINT CHARGE DEFECT

The crystalline lattice defects have strong influence on all phenomena associated with domains. This relates both to statics and dynamics of the domain structure. To describe these phenomena, it is necessary first of all to consider the processes of interaction of the domain boundaries with defects and determine the force or energy characteristics of these interactions.

In the process of interaction with defects there is a possibility of changes in the profile of the ferroelectric or ferroelastic domain wall. Some of these changes can result in formation of bound electric charges or twinning dislocations in the areas of the bent domain wall. The formation of the charges (twinning dislocations) results in the appearance of long-range electrical or elastic fields. To calculate these fields, the equation of motion (or equilibrium) of the domain wall should be supplemented by the equations of electrodynamics or elasticity theory.

The general topic of this chapter in this section starts with consideration of interaction 180° domain walls in ferroelectrics with point charged defects. The equilibrium equation of the domain wall represents the condition of equality to zero of the total pressure onto any section of the boundary. In this case it has the following form

$$-\gamma\left(\frac{\partial^2 U}{\partial z^2}+\frac{\partial^2 U}{\partial y^2}\right)+2P_0\left.\frac{\partial \varphi}{\partial z}\right|_{x=0}=0. \qquad (1.1)$$

The first term in (1.1) is the pressure on the boundary related to the increase of the surface of the boundary due to its bending

– the so-called Laplace pressure [95], the second term is the pressure from the direction of the electric field in the system under consideration, described by electrostatic potential $\varphi$. Hereinafter for linearization of the discussed ratios the values of the calculated fields are considered not in the actual positions of the deflected boundary but in the area of its nondisplaced position. Evidently, the approximation used will become more accurate the less the boundary is deflected. The boundaries of permissibility of approximation made will be discussed directly in the specific cases. In the problem under consideration, the position of the nondisplaced boundary coincides with coordinate $zy$-plane and the polar direction with the $z$-axis.

The equation, supplementing (1.1), is the electrostatic equation $\mathrm{div}\mathbf{D} = 0$, where $\mathbf{D}$ is the vector of electrostatic induction. Discriminating between spontaneous and induced polarization of a crystal, we can write

$$\mathrm{div}\,\mathbf{D} = \mathrm{div}\,\varepsilon_{ij}E_j + 4\pi\,\mathrm{div}\,\mathbf{P}_s = 4\pi\rho. \tag{1.2}$$

Here $\varepsilon_{ij}$ is the tensor of dielectric permittivity of the monodomain crystal, $\rho = Ze\cdot\delta(\mathbf{r}-\mathbf{r}_d)$ is the volume density of the charge corresponding to a single charge with the value $Ze$, $\mathbf{P}_s$ is the distribution of spontaneous polarization in the ferroelectric in the presence of a domain structure. In a crystal with a single wall

$$\mathbf{P}_s = -\mathbf{P}_0\left[1 - 2\Theta(x-U)\right], \tag{1.3}$$

where $\theta(x)$ is the Heaviside function. When writing (1.3) to simplify the equation, the approximation of a structureless boundary (Fig.4.1) is used.

Placing in the limit of small boundary displacements the bound electric charges formed on it in the process of bending into the

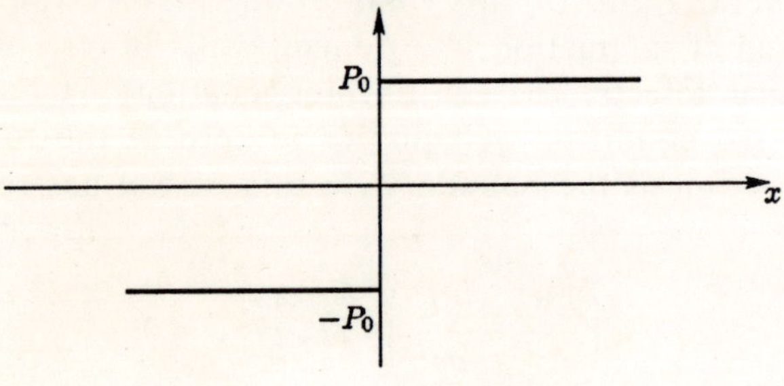


**Fig.4.1.** Distribution of polarization in a crystal with a single unbent wall in the model of a structureless boundary.

plane of nondisplaced boundary, we obtain from (1.3)

$$\operatorname{div}\mathbf{P}_s=\frac{\partial P}{\partial z}=2P_0\frac{\partial\Theta}{\partial U}\cdot\frac{\partial U}{\partial z}\simeq 2P_0\delta(x)\frac{\partial U}{\partial z}. \tag{1.4}$$

Substituting now (1.4) into (1.2) and supplementing the resultant ratio by the boundary equilibrium equation for $\varepsilon_{ij}=\varepsilon_i\cdot\delta_{ij}$ ($\varepsilon_i=\varepsilon_c$ or $\varepsilon_a$ for $i=1$ or $i=2,3$, respectively) taking into account the ratio $E_j=-\partial\varphi/\partial x_j$ we obtain the following simultaneous equations, that determine the profile of the boundary bent in the field of the extraneous charge $Ze$

$$\begin{gathered}\varepsilon_c\frac{\partial^2\varphi}{\partial z^2}+\varepsilon_a\left(\frac{\partial^2\varphi}{\partial y^2}+\frac{\partial^2\varphi}{\partial x^2}\right)=\\ =8\pi P_0\cdot\delta(x)\cdot\frac{\partial U}{\partial z}-4\pi Ze\cdot\delta(\mathbf{r}-\mathbf{r}_d),\\ -\gamma\left(\frac{\partial^2U}{\partial z^2}+\frac{\partial^2U}{\partial y^2}\right)+2P_0\left.\frac{\partial\varphi}{\partial z}\right|_{x=0}=0.\end{gathered} \tag{1.5}$$

The radius-vector $\mathbf{r}_d$ describes here the position of the defect.

Let us use the expansion into two-dimensional Fourier integrals for displacement of the boundary and electrostatic potential:

$$\begin{gathered}U(y,z)=\int U_{\vec{k}}\cdot\exp(i\mathbf{kp})\frac{dk}{(2\pi)^2},\\ \varphi(x,y,z)=\int\varphi_{\vec{k}}(x)\cdot\exp(i\mathbf{kp})\frac{d\mathbf{k}}{(2\pi)^2},\\ \rho=(y,z).\end{gathered} \tag{1.6}$$

Substituting expansions (1.6) into the first of the equations in (1.5) enables us to write it in the following form:

$$\varphi_{\mathbf{k}}''-\left(\frac{\varepsilon_c}{\varepsilon_a}k_z^2+k_y^2\right)\varphi_{\mathbf{k}}=\frac{8\pi P_0\cdot ik_z}{\varepsilon_a}U_{\mathbf{k}}\delta(x)-\frac{4\pi Ze}{\varepsilon_a}\delta(x-x_d), \tag{1.7}$$

where it is assumed that the defect is located at a point with radius vector $\mathbf{r}_d=(x_d,0,0)$ The solution of equation (1.7) using, for example, another Fourier expansion now for $\varphi_{\mathbf{k}}(x)$ in respect of $x$, yields:

$$\varphi_{\mathbf{k}}(x) = -\frac{4\pi P_0 i k_z U_{\mathbf{k}}}{\varepsilon_a \sqrt{\frac{\varepsilon_c}{\varepsilon_a} k_z^2 + k_y^2}} \cdot \exp\left(-|x|\sqrt{\frac{\varepsilon_c}{\varepsilon_a} k_z^2 + k_y^2}\right) +$$

$$+\frac{2\pi Ze}{\varepsilon_a \sqrt{\frac{\varepsilon_c}{\varepsilon_a} k_z^2 + k_y^2}} \cdot \exp\left(-|x - x_d|\sqrt{\frac{\varepsilon_c}{\varepsilon_a} k_z^2 + k_y^2}\right). \qquad (1.8)$$

According to the second of the equations in set (1.5)

$$\varphi_{\bar{k}}(x=0) = -\frac{\gamma k^2}{2P_0 i k_z}, \quad k^2 = k_z^2 + k_y^2. \qquad (1.9)$$

Equating equation (1.8) taken at $x = 0$ and ratio (1.9) we obtain an equation for determination of the Fourier coefficient of the boundary displacement $U_{\mathbf{k}}$ where from

$$U_{\bar{k}} = \frac{-4\pi P_0 Ze \cdot ik_z \cdot \exp\left(-|x_d|\sqrt{\frac{\varepsilon_c}{\varepsilon_a} k_z^2 + k_y^2}\right)}{\left[8\pi P_0^2 k_z^2 + \gamma k^2 \varepsilon_a \sqrt{\frac{\varepsilon_c}{\varepsilon_a} k_z^2 + k_y^2}\right]}. \qquad (1.10)$$

Substitution of (1.10) into equation (1.8) gives the final expression for the Fourier expansion of the potential [96, 97]:

$$\varphi_{\bar{k}} = \frac{2\pi Ze}{\varepsilon_a \sqrt{\frac{\varepsilon_c}{\varepsilon_a} k_z^2 + k_y^2}} \cdot \exp\left(-|x - x_d|\sqrt{\frac{\varepsilon_c}{\varepsilon_a} k_z^2 + k_y^2}\right) -$$

$$-\frac{(4\pi P_0)^2 k_z^2 \cdot Ze}{\varepsilon_a \sqrt{\frac{\varepsilon_c}{\varepsilon_a} k_z^2 + k_y^2}} \cdot \frac{\exp\left(-(|x_d| + |x|)\sqrt{\frac{\varepsilon_c}{\varepsilon_a} k_z^2 + k_y^2}\right)}{\left[8\pi P_0^2 k_z^2 + \gamma k^2 \varepsilon_a \sqrt{\frac{\varepsilon_c}{\varepsilon_a} k_z^2 + k_y^2}\right]}. \qquad (1.11)$$

The first term in (1.11) is a two-dimensional Fourier image of the point charge potential. The second is the potential of the bound charges on the bent domain wall induced by the point charge. Integration of the latter makes it possible to find the coordinate dependence of the induced potential.

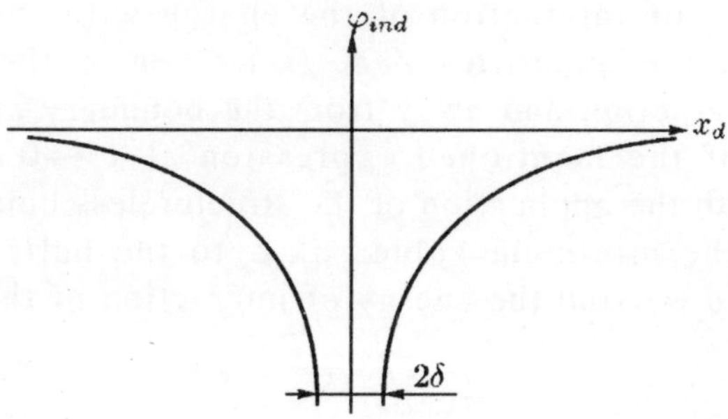


**Fig 4.2.** Variation of the electrostatic potential, induced by the bent wall in the area of location of the charge defect, in relation to the distance between them (linear approximation).

Neglecting surface tension, the dependence of this potential, taken in the area of the defect location, on the distance from the defect to the initial position of the boundary, has the form (Fig.4.2):

$$\varphi_{ind} = -\frac{Ze}{\sqrt{\varepsilon_c \varepsilon_a} \cdot 2|x_d|}. \qquad (1.12)$$

The obtained result has a clear physical interpretation. As we can see in Fig.4.3, the antisymmetry of the field of the charged defect along the polar axis with respect to the perpendicular to the boundary, passing though the defect, results in unlike signs of the pressure on the boundary from the direction of this field at $z>0$ and $z<0$. Moving in different directions under the influence of this field the boundary finally bends in the vicinity of point $z = 0$. At that the direction of the bending is always such that the sign of the bound charge occurring at the bent boundary is always opposite to the sign of the charged defect. The attraction of the charges of unlike signs means that the interaction of the charged

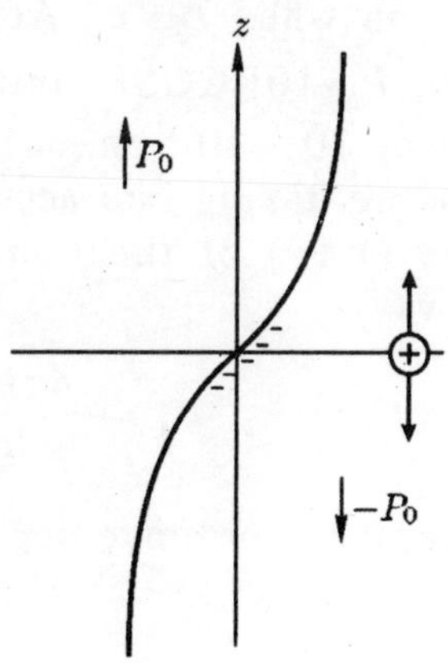


**Fig.4.3.** Displacement of a domain wall under the influence of the field of a charged defect.

defect of any sign with the domain wall is always of the attraction type.

The energy of interaction of the charge with the wall is the difference of the quantities $Ze\varphi_{ind}(x_d)$ taken in the area of the maximum interaction and away from the boundary at $x_d \to \infty$. The divergence of the mentioned expression at $x_d \to 0$ is obviously associated with the application of the structureless boundary model. Restricting the minimum values of $x_d$ to the half width of the domain wall $\delta$ we find the energy of interaction of the defect with the boundary

$$\mathcal{U}_0 = \frac{Z^2 e^2}{2\sqrt{\varepsilon_c \varepsilon_a} \cdot \delta}. \tag{1.13}$$

For $Z = 1$, $\varepsilon_c \sim 10^3$, $\varepsilon_a \sim 10$, $\delta \sim 10^{-7}$ cm the value of $\mathcal{U}_0$ is of the order of $10^{-3}$ eV, and at $\varepsilon_c \varepsilon_a \sim 10^2$ it is an order of magnitude higher.

It should be mentioned that equation (1.13) for $\mathcal{U}_0$ can be written immediately if it is taken into account that the interaction of the charged defect with the bound charge induced by this defect on the boundary is in fact the interaction with an image charge, and the domain wall itself in accordance with (1.1) ignoring the surface tension is the equipotential surface. The flow of the bound charges on this surface as a result of a bending of its profile is similar to the motion of free charges on a metallic surface.

When determining the specific form of the bending of the boundary let us consider the case of $x_d = 0$ and the most typical situation when $\varepsilon_c \gg \varepsilon_a$. At conventional $\gamma \sim 0.1 \div 1$ erg/cm$^2$, $\varepsilon_c \sim 10^3$, $\varepsilon_a \sim 10$, $P_0 \sim 10^4$ CGSE units, the ratio $\lambda = \gamma\sqrt{\varepsilon_c \varepsilon_a}/8\pi P_0^2$ is of the order of $10^{-8} \div 10^{-7}$ cm, whereas the maximum is $k_z \sim 2\pi/\delta \sim 10^7$cm$^{-1}$. Therefore, taking into account the smallness of $k_z\lambda \leq 1$, the Fourier image (1.10) of the boundary displacement can be written as follows

$$U_{\mathbf{k}} = -\frac{4\pi P_0 Z e}{\gamma\sqrt{\varepsilon_c \varepsilon_a}} \cdot \frac{ik_z}{\left(k_y^2|k_z| + k_z^2/\lambda\right)}. \tag{1.14}$$

Hence, the coordinate dependence of the boundary displacement is

$$U(y,z)=\frac{Ze}{2\pi P_0\sqrt{\lambda z}}\sqrt{\frac{\pi}{2}}\left[\frac{1}{2}-C\left(\frac{p}{\sqrt{2\pi}}\right)\right]\cos\left(\frac{p^2}{4}\right)+$$
$$+\left[\frac{1}{2}-S\left(\frac{p}{\sqrt{2\pi}}\right)\right]\sin\left(\frac{p^2}{4}\right), \tag{1.15}$$
$$p=y/\sqrt{\lambda z},$$

where $C(x)$, $S(x)$ are Frenel's integrals. In order to analyse the expression in the braces (1.15) it can be conveniently approximated by the polynomial [98]

$$\left[\frac{1}{2}-(T)\right]\cos\left(\frac{\pi}{2}T^2\right)+\left[\frac{1}{2}-S(T)\right]\sin\left(\frac{\pi}{2}T^2\right)=g(T), \tag{1.16}$$

where

$$g(T)=\frac{1}{2+P\cdot T+Q\cdot T^2+R\cdot T^3},$$
$$T=\frac{p}{\sqrt{2\pi}}=\frac{y}{\sqrt{2\pi\lambda z}},\quad T>0. \tag{1.17}$$

Here $p = 4.142$, $Q = 3.492$, $R = 6.670$ [98].

Using (1.17), it is convenient to write the boundary displacement (1.15) in the following form

$$U(y,z)=\frac{Ze}{2P_0}\times$$
$$\times\frac{2\pi\lambda z}{\left[2(2\pi\lambda|z|)^{\frac{3}{2}}+p|y|\cdot 2\pi\lambda|z|+Qy^2\sqrt{2\pi\lambda|z|}+R|y|^3\right]}. \tag{1.18}$$

As expected, the displacement (1.18) is asymmetrical along the polar axis $z$ with respect to the position of the charged defect, and possesses the characteristic law of decrease $1/\sqrt{z}$ along the polar axis and dependence $U\sim 1/y^3$ in the perpendicular direction. With increasing $z$, the displacement of the boundary, remaining maximum for $y = 0$, spreads along the $y$-direction decreasing in value simultaneously. At the same time, the integral from $U(y,z)dy$ remains equal to a constant (otherwise, we would be faced with the localization of the bound charge at the boundary in the vicinity of the fixing point, i.e. not with the compensation of the point defect by the bound charge at the boundary, but only with the redistribution of the density of the latter with its general zero value). This is very well illustrated, for example, when considering the Fourier image of the boundary displacement. Integral from $U(y,z)$ over $dy$ yields delta function $\delta(k_y)$. Then assuming that $k_y = 0$ in (1.14), we obtain $\int U(y,z)dy = \frac{Ze}{2P_0}\Theta(z)$, where $\Theta(z)$ is the sign function of $z$, i.e. we obtain a constant value at any given $z$ and independent of the coefficient of the boundary surface tension.

For a 90° domain wall, as it was shown in section 2.2, the interaction of the charged carriers with the domain wall takes place not only by way of distortion of its profile similar to the one discussed above, but also by way of the interaction with an internal electric field existing in such a boundary.

## 4.2. Dislocation description of bent domain walls in ferroelastics. Equation of incompatibility for spontaneous deformation

As it was already mentioned in chapter 1, the domains in ferroelastics are mechanical twins that differ in the simplest case in the sign of spontaneous shear deformation, and the plane of the domain boundary at that coincides, as a rule, with the so-called invariant plane, i.e. the plane in which the positions of the atoms remain unchanged in the process of rearrangement of the crystalline structure during a phase transition. At each point of such a boundary, the deformations are compatible and the continuity of the medium is preserved (Fig.4.4 a).

When the domain wall deviates from the invariant plane, there are breaks in continuity which, depending on the direction of displacement of the wall (Fig.4.4 b, c), can be described by twinning edge (Fig.4.4 b) or screw (Fig.4.4 c) dislocations with Burger's vectors **b** considerably smaller than the lattice constant $a$. At that any macroscopic inclination of the domain wall can be ensured using the appropriate set of twinning dislocations. For small inclinations of the domain wall the mentioned dislocations can be assumed as located in the initial invariant plane. The change of the angle of inclination of the domain wall in this description is associated with the variation of the density of twinning dislocations and with their gliding in the invariant plane.

Let us give a mathematical description of an inclined domain wall in a ferroelastic and use toward this end the main ratios of the dislocation theory of elasticity. According to the initial definition of

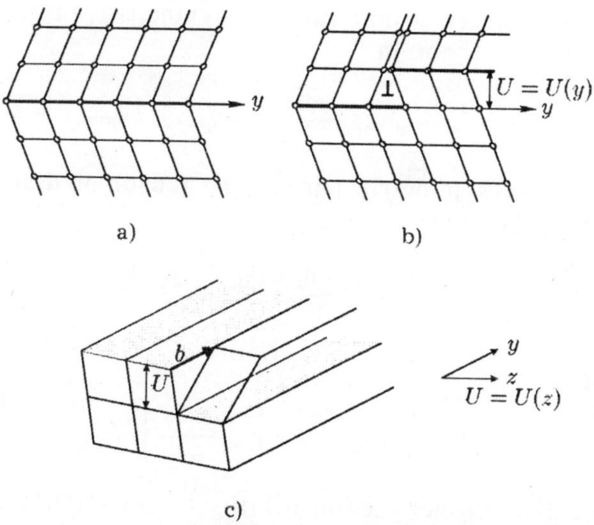


**Fig.4.4.** Different orientations of the domain wall in ferroelastics. (a) the domain boundary coincides with the invariant plane, (b) the inclined wall, the displacement of the wall depends on the coordinate in the direction of spontaneous shear, (c) the inclined wall, its displacement changes in the direction perpendicular to spontaneous shear.

the dislocations [99-101], when traversing any closed contour, enveloping a set of dislocation lines, the vector of elastic displacement of the medium **u** gets a finite increment equal to the sum of the Burgers vectors **b** of all the dislocation lines enclosed in this contour. This definition can be written in the following form

$$\oint_e du_k = \oint_l \frac{\partial u_k}{\partial x_i} dx_i = -b_k, \tag{2.1}$$

where the tensor $u_{ik}=\partial u_k/\partial x_i$ is the tensor of elastic distortion. Substituting in (2.1) the contour integral by the integral of the surface resting on the contour we obtain

$$\oint_l u_{ln} \cdot dx_l = \int d\sum_i e_{ikl} \frac{\partial u_{ln}}{\partial x_k} \tag{2.2}$$

and introduçing the tensor of dislocation density $\alpha_{in}$, with the help of ratio (2.2) instead of (2.1) we obtain a different ratio

$$e_{ikl} \frac{\partial u_{ln}}{\partial x_k} = -\alpha_{in}, \tag{2.3}$$

where $e_{ikl}$ is the unit antisymmetric tensor. Using differential operation $e_{jmn}\partial/\partial x_m$ once more for the both parts of (2.3) and ensuring that the resultant ratio is symmetric in respect of the indices $i$ and $j$, we obtain

$$-e_{ikl}\cdot e_{jmn}\frac{\partial^2\varepsilon_{ln}}{\partial x_k\partial x_m}=\eta_{ij}, \tag{2.4}$$

where $\varepsilon_{ln}$ is the symmetric part of the tensor of distortion or, in other words, the strain tensor

$$\varepsilon_{ln}=\frac{1}{2}\left(\frac{\partial u_l}{\partial x_n}+\frac{\partial u_n}{\partial x_l}\right), \tag{2.5}$$

and

$$\eta_{ij}=\frac{1}{2}\left(e_{jmn}\frac{\partial\alpha_{in}}{\partial x_m}+e_{imn}\cdot\frac{\partial\alpha_{jn}}{\partial x_m}\right) \tag{2.6}$$

is the so-called Kröner incompatibility tensor [102,103]. The name of the latter is related with the fact that in the absence of bending of the domain wall, i.e. in the absence of twinning dislocations, $\eta_{ij}=0$ and ratio (2.4) is transformed into the well-known condition of compatibility of strains – the St-Venant condition [102]:

$$e_{ikl}\cdot e_{jmn}\frac{\partial^2\varepsilon_{ln}}{\partial x_k\partial x_m}=0. \tag{2.7}$$

Ratio (2.4) makes it possible to determine the distribution of elastic strains from the known tensor of incompatibility. It can be transformed to the equation for stresses using Hooke's law. To carry out this operation, let us rewrite equation (2.4) excluding from it, with the help of the ratio

$$e_{ijk}\cdot e_{klm}=\delta_{il}\cdot\delta_{jm}-\delta_{im}\cdot\delta_{jl} \tag{2.8}$$

the unit antisymmetric tensor. Consequently we obtain [102]

$$\begin{aligned}&-e_{ikl}\cdot e_{jmn}\cdot\varepsilon_{ln,km}=-e_{ikl}\cdot e_{jmn}\cdot\varepsilon_{pq,km}\cdot\delta_{pl}\cdot\delta_{qn}=\\&=-e_{ikl}\cdot e_{jmn}\cdot\varepsilon_{pq,km}\cdot\left(e_{npr}\cdot e_{rql}+\delta_{pq}\cdot\delta_{ln}\right)=-\\&-\left(\delta_{ir}\cdot\delta_{kq}-\delta_{iq}\cdot\delta_{kr}\right)\cdot\left(\delta_{jp}\cdot\delta_{mr}-\delta_{jr}\cdot\delta_{mp}\right)\cdot\varepsilon_{pq,km}-\\&-\left(\delta_{ij}\cdot\delta_{km}-\delta_{im}\cdot\delta_{kj}\right)\cdot\varepsilon_{pp,km}=\varepsilon_{ji,kk}+\varepsilon_{kk,ij}-\\&-\left(\varepsilon_{jk,ki}+\varepsilon_{ki,jk}\right)+\left(\varepsilon_{kl,kl}-\varepsilon_{kk,ll}\right)\cdot\delta_{ij}=\eta_{ij}.\end{aligned} \tag{2.9}$$

where $\delta_{ij}$ is the Kronecker symbol, and the indices after the comma indicate the differentiation in respect of the corresponding coordinate.

Substitution of Hooke's law into (2.9), written for the isotropic case

$$\varepsilon_{ij} = \frac{1}{2\mu}\left(\sigma_{ij} - \frac{1}{(m+1)}\sigma_{kk}\cdot\delta_{ij}\right), \quad m = 2(\lambda + m)/\lambda, \tag{2.10}$$

where λ and μ are the Lame coefficients gives

$$\begin{gathered}\sigma_{ij,kk} + \frac{m}{m+1}\left(\sigma_{kk,ij} - \sigma_{kk,ll}\cdot\delta_{ij}\right) - \\ -\left(\sigma_{jk,ki} + \sigma_{ki,jk} + \sigma_{kl,kl}\cdot\delta_{ij}\right) = 2\mu\cdot\eta_{ij}.\end{gathered} \tag{2.11}$$

Using the equation of dynamics of the elastic medium

$$\sigma_{ij,j} + f_i = \rho\cdot\ddot{u}_i \tag{2.12}$$

where $f_i$ is the corresponding projection of the volume density of the external forces, $\rho$ is the density of the medium, instead of (2.11) we obtain

$$\begin{gathered}\sigma_{ij,kk} + \frac{m}{m+1}\left(\sigma_{kk,ij} - \sigma_{kk,ll}\cdot\delta_{ij}\right) - \\ -\left[\left(\rho\ddot{u}_j\right)_{,i} + \left(\rho\ddot{u}_i\right)_{,j}\right] + \left(\rho\ddot{u}_l\right)_{,l}\cdot\delta_{ij} = 2\mu\cdot\eta_{ij}.\end{gathered} \tag{2.13}$$

Taking now into account that the quantities $u_{ij}$ in (2.13) are in the general case the sum of the spontaneous $u_{ij}^s$ and elastic distortion, and introducing the tensor of the density of the flow of dislocations

$$j_{ij} = -\frac{\partial u_{ij}^s}{\partial t}, \tag{2.14}$$

using the ratios (2.5) and (2.10) for the elastic part of the tensor $u_{ij}$ and the ratio (2.14) for its inelastic part, equation (2.13) can be written in the final form

$$\begin{gathered}\sigma_{ij,kk} + \frac{m}{m+1}\left(\sigma_{kk,ij} - \sigma_{kk,ll}\cdot\delta_{ij}\right) - \frac{\rho}{\mu}\ddot{\sigma}_{ij} + \\ + \frac{\rho}{\mu}\cdot\frac{(\lambda+\mu)}{(3\lambda+2\mu)}\cdot\ddot{\sigma}_{kk}\delta_{ij} + f_{j,i} + f_{i,j} - f_{l,l}\cdot\delta_{ij} + \\ + \rho\frac{\partial}{\partial t}\left(j_{ij} + j_{ji}\right) - \rho\frac{\partial}{\partial t}j_{ll}\cdot\delta_{ij} = 2\mu\cdot\eta_{ij}.\end{gathered} \tag{2.15}$$

The resultant equation is referred to as the Beltrami–Mitchell dynamic equation [102]. It enables using the available sources of fields, described by the tensor $\eta_{ij}$ and $f_i$, to determine the elastic stresses caused by them. Thus, in the elasticity theory this equation plays the role identical to that of Maxwell's equations in electrodynamics.

## 4.3. Interaction of the ferroelectric-ferroelastic domain boundary with a point charged defect

In ferroelectric–ferroelastic crystals, the phase transition to the polar state is accompanied by the occurrence of spontaneous deformation. In this case the bending of the boundary during its interaction with a defect results in the appearance of not only bound electric charges but also of twin dislocation in the boundary plane. Evidently, the latter will also influence the nature of boundary bending and consequently the energy of the boundary interaction with the defect.

Let us consider now the interaction of the boundary with a charged defect in a ferroelectric–ferroelastic using crystals with the symmetry of potassium dihydrophosphate as an example, in which the formation of polarization along $z$-axis is accompanied by the appearance of spontaneous shear deformation in the perpendicular plane $\varepsilon_{12} \equiv \epsilon_0$.

The set of equations describing this interaction has the form

$$\varepsilon_c \frac{\partial^2 \varphi}{\partial z^2} + \varepsilon_a \left( \frac{\partial^2 \varphi}{\partial y^2} + \frac{\partial^2 \varphi}{\partial x^2} \right) = \\ = 8\pi P_0 \delta(x) \cdot \frac{\partial U}{\partial z} - 4\pi Ze \cdot \delta(\mathbf{r} - \mathbf{r}_d), \\ -2P_0 \frac{\partial \varphi}{\partial z}\Big|_{x=0} + 2\epsilon_0 \sigma_{12}\Big|_{x=0} = 0. \tag{3.1}$$

As it can be seen from (3.1), in this case the equation of the boundary equilibrium also includes the term related to the pressure of the boundary from the direction of the field of elastic stresses (it is written in the form similar to the pressure from the direction of the electric field). At the same time, in the equation of the boundary equilibrium the surface tension is ignored, which as it will be shown later, is considerably smaller here than the other terms for all orientations of bending and all values of the wave vector $k$.

For the combined solution of equations of set (3.1) it is necessary to find first of all the relation of the stresses, formed at the bending of the boundary, to the magnitude and orientation of its bending. In the static situation discussed here the stresses accompanying the bending of the boundary are found from the static Beltrami equation in the absence of the external elastic forces. Taking into account the equality $f_i$, $j_{ij} = 0$ and the absence of time dependence of the $\sigma_{ij}$ values the latter has the form

$$\sigma_{ij,kk} + \frac{m}{m+1}\left(\sigma_{kk,ij} - \sigma_{kk,ll} \cdot \delta_{ij}\right) = 2\mu\eta_{ij}. \tag{3.2}$$

Let us find the components of the tensor of incompatibility unequal zero and their relation to the bending of the boundary. The distribution of spontaneous distortion in the crystal with a single domain wall, coinciding in its initial state with plane $zy$, is:

$$u_{12}^{s} = -\epsilon_0\left[1 - 2\Theta\left(x - U\left(z, y\right)\right)\right]. \tag{3.3}$$

The bends of the boundary in the direction of spontaneous shear (along axis $y$) and in perpendicular direction result in the formation of edge and screw dislocations [104–107], distributed in accordance with (2.3) with the densities

$$\begin{aligned}
\alpha_{22} &= e_{231}\frac{\partial u_{12}^{s}}{\partial x_3} = 2\epsilon_0\frac{\partial U}{\partial z}\delta(x), \\
\alpha_{32} &= -e_{321}\frac{\partial u_{12}^{s}}{\partial x_2} = -2\epsilon_0\frac{\partial U}{\partial y}\delta(x).
\end{aligned} \tag{3.4}$$

Substituting (3.4) into (2.6) we obtain components of the tensor of incompatibility differing from zero

$$\begin{aligned}
\eta_{12} &= -\frac{1}{2}\frac{\partial \alpha_{22}}{\partial x_3} = -\epsilon_0\frac{\partial^2 U}{\partial z^2}\delta(x), \\
\eta_{13} &= -\frac{1}{2}\frac{\partial \alpha_{32}}{\partial x_3} = \epsilon_0\frac{\partial^2 U}{\partial y \partial z}\delta(x), \\
\eta_{33} &= -\frac{\partial \alpha_{32}}{\partial x_1} = -2\epsilon_0\frac{\partial U}{\partial y}\delta'(x), \\
\eta_{23} &= \frac{1}{2}\frac{\partial \alpha_{22}}{\partial x_1} = \epsilon_0\frac{\partial U}{\partial z}\delta'(x).
\end{aligned} \tag{3.5}$$

Let us use the two-dimensional Fourier expansion for the solution of equation (3.2)

$$U(y,z)=\int U_{\mathbf{k}}\cdot\exp(i\mathbf{kp})\frac{d\mathbf{k}}{(2\pi)^2},$$
$$\sigma_{ij}=\int\tilde{\sigma}_{ij}(x)\cdot\exp(i\mathbf{kp})\frac{d\mathbf{k}}{(2\pi)^2},\quad \mathbf{\rho}=(y,z). \tag{3.6}$$

Consequently, on the basis of (3.2) and (3.6) the set of equations for the Fourier image $\tilde{\sigma}_{11},\tilde{\sigma}_{22},\tilde{\sigma}_{33}$ has the form

$$\tilde{\sigma}_{11}''-k^2\tilde{\sigma}_{11}+\beta k^2\tilde{\sigma}=0,\quad \tilde{\sigma}_{22}''-k^2\tilde{\sigma}_{22}+\beta\left(\tilde{\sigma}\cdot k_z^2-\tilde{\sigma}''\right)=0,$$
$$\tilde{\sigma}_{33}''-k^2\tilde{\sigma}_{33}+\beta\left(\tilde{\sigma}\cdot k_y^2-\tilde{\sigma}''\right)=-4\mu\varepsilon_0 ik_yU_{\mathbf{k}}\cdot\delta'(x), \tag{3.7}$$

where $\beta=m/(m+1)$, $k^2=k_y^2+k_z^2$, $\tilde{\sigma}=\tilde{\sigma}_{11}+\tilde{\sigma}_{22}+\tilde{\sigma}_{33}$.

Adding up equations (3.7), we obtain an equation for determining $\tilde{\sigma}$:

$$\tilde{\sigma}''-k^2\tilde{\sigma}=\frac{4\mu\epsilon_0 ik_yU_{\mathbf{k}}\cdot\delta'(x)}{(2\beta-1)}. \tag{3.8}$$

Let us use again the Fourier expansion:

$$\tilde{\sigma}(x)=\int_{-\infty}^{\infty}\sigma_{k_x}\cdot\exp(ik_xx)\frac{dk_x}{2\pi}. \tag{3.9}$$

Substituting (3.9) into (3.8) and solving the resultant equation in relation to $\sigma_{k_x}$, we obtain

$$\sigma_{k_x}=\frac{4\mu\epsilon_0 k_xk_yU_{\mathbf{k}}}{(2\beta-1)\left(k_x^2+k_y^2\right)}. \tag{3.10}$$

Whence

$$\tilde{\sigma}(x)=\frac{2\mu\epsilon_0 ik_yU_{\mathbf{k}}}{(2\beta-1)}\cdot\exp(-|x|k)\cdot\operatorname{sign}x. \tag{3.11}$$

On the basis of (3.2) and (3.5) the equation for determining the Fourier image $\tilde{\sigma}_{12}(x)$ has the form

$$\tilde{\sigma}_{12}''-k^2\tilde{\sigma}_{12}+\beta ik_y\tilde{\sigma}'=2\mu\epsilon_0 k_z^2\delta(x)\cdot U_{\mathbf{k}}. \tag{3.12}$$

Using expansion (3.9) for $\tilde{\sigma}(x)$ and the identical expansion for $\tilde{\sigma}_{12}(x)$, on the basis of (3.12) we obtain

$$\sigma_{12}^{k_x}=-\frac{2\mu\epsilon_0 k_z^2}{\left(k_x^2+k^2\right)}U_{\mathbf{k}}-\frac{\beta k_yk_x}{\left(k_x^2+k^2\right)}\sigma_{k_x}. \tag{3.13}$$

Or taking into account (3.10)

$$\sigma_{12}^{k_x} = -\frac{2\mu\epsilon_0 k_z^2}{\left(k_x^2 + k^2\right)} U_{\mathbf{k}} - \frac{4\mu\epsilon_0 \beta k_y^2 k_x^2}{(2\beta-1)\left(k_x^2+k^2\right)^2} U_{\mathbf{k}}. \tag{3.14}$$

Hence

$$\tilde{\sigma}_{12}(x) = -\left\{\frac{\mu\epsilon_0 k_z^2}{k} U_{\mathbf{k}} + \frac{\mu\epsilon_0\beta}{(2\beta-1)} \cdot \frac{k_y^2}{k} \cdot U_{\mathbf{k}}\left(1-k|x|\right)\right\} \cdot \exp(-k|x|). \tag{3.15}$$

and, therefore

$$\tilde{\sigma}_{12}(x=0) = -\left\{\frac{\mu\epsilon_0}{k} U_{\mathbf{k}}\left(k_z^2 + \omega k_y^2\right)\right\},$$

$$\omega = \frac{\beta}{2\beta-1} = \frac{m}{m-1} = \frac{2(\lambda+\mu)}{\lambda+2\mu}. \tag{3.16}$$

Now using the solution of the first of the equations of system (3.1) in the form of (1.6), (1.8), found in section 4.1, for the present case of the ferroelectric-ferroelastic crystal we obtain the following Fourier image of the displacement of the boundary in the field of the point charge defect

$$U_{\mathbf{k}} = \frac{-4\pi P_0 \cdot Zeik_z \cdot \exp\left(-|x_d|\sqrt{\frac{\varepsilon_c}{\varepsilon_a}k_z^2 + k_y^2}\right)}{\varepsilon_a\sqrt{\frac{\varepsilon_c}{\varepsilon_a}k_z^2 + k_y^2} \cdot \left\{\frac{8\pi P_0 k_z^2}{\varepsilon_a\sqrt{\frac{\varepsilon_c}{\varepsilon_a}k_z^2 + k_y^2}} + \frac{2\mu\epsilon_0^2}{k}\left(k_z^2 + \omega k_y^2\right)\right\}}. \tag{3.17}$$

As in the case for a 'pure' ferroelectric, the energy of interaction of the defect with the boundary here is the difference of the values at $Ze\cdot\varphi_{ind}(xd)$ in the area of the maximum interaction and away from the boundary. Taking into account (1.6), (1.8) and the expression derived here for $U_k$ (3.17), the interaction energy in the case of $\varepsilon_c = \varepsilon_a \equiv \varepsilon$ is equal to

$$\mathcal{U}_0 = \frac{Z^2 e^2}{2\varepsilon\delta} \cdot \frac{\gamma}{\left[(1+\gamma) + \sqrt{1+\gamma}\cdot\sqrt{\omega}\right]}, \quad \gamma = \frac{4\pi P_0^2}{\varepsilon\mu\epsilon_0^2},$$

$$\omega = 2(\lambda+\mu)/(\lambda+2\mu) \tag{3.18}$$

As can be seen from (3.18), at $\epsilon_0 \to 0$, i.e. at $\gamma \to \infty$ the energy of interaction of the defect with the boundary is determined by electrostatics only and converts to the equation (1.13). At values of the coefficient $\gamma$, that differ from zero (coefficient $\gamma$ describes the relative role of the electrical and elastic interaction controlling the displăcement of the boundary), the additional rigidity, preventing the displacement of the boundary, and related to the appearance of the elastic fields at the bending of the domain wall of the ferroelectric–ferroelastic, results in a decrease of the energy of interaction of the boundary with the defect.

The calculation of the boundary displacement on the basis of equation (3.17) for the case, in which the defect is located directly on the boundary, gives at $\varepsilon_c=\varepsilon_a\equiv\varepsilon$ the displacement

$$U(y,z)=\frac{ZeP_0}{\sqrt{1+\gamma}\cdot\sqrt{\omega}\cdot\mu\varepsilon_0^2\cdot\varepsilon}\cdot\frac{z}{\left(z^2+\dfrac{\gamma+1}{\omega}y^2\right)}, \tag{3.19}$$

which like in (1.18) is asymmetric along the polar axis $z$.

## 4.4. Interaction of the domain boundary in ferroelastic with a dilatation centre

The defects of the crystalline lattice – internodal atoms and vacancies as well as the impurities introduced into the crystal cause the deformation of the lattice of a specific sign in its nearest environment thus creating round themselves a certain distribution of stresses. Similarly to interaction of the charge defect with the domain boundary in the ferroelectric a certain part of the stresses generated by an external source, i.e. by the defect can be relieved of the domain boundary by its bending. This makes the position of the domain wall in the ferroelastic in the vicinity of the defect generating elastic fields more energy advantageous in comparison with their isolated distribution, i.e. results in the interaction of the boundary with the defect.

The simplest model of the point defect in the elasticity theory is the so-called dilatation centre whose influence on the nearest environment is equivalent to the influence of three pairs of equal forces applied to the location of the defect and directed along the coordinate axes. In the elasticity theory, this defect is described by the volume density of forces of the following type:

$$f(\mathbf{r}) = -\left(\lambda + \frac{2}{3}\mu\right) \cdot \Omega_0 \cdot \text{grad}\,\delta(\mathbf{r} - \mathbf{r}_d), \tag{4.1}$$

where $\mathbf{r}_d$ is the coordinate of the defect. In the cubic crystal or in the isotropic medium, $\Omega_0$ has a simple physical meaning. Its value is equal to the change of the crystal volume caused by the presence of a single defect in the crystal. For an internodal atom $\Omega_0>0$ and for a vacancy, where displacement of the adjacent atoms takes place in the direction of the defect, $\Omega_0<0$.

The simultaneous equations describing the interaction between the centre of dilatation and the domain boundary in a ferroelastic are represented by the following set of equations

$$\begin{aligned} &\sigma_{ij,kk} + \frac{m}{m+1}\left(\sigma_{kk,ij} - \sigma_{kk,ll} \cdot \delta_{ij}\right) + \\ &\qquad\qquad + f_{i,j} + f_{j,i} - f_{l,l} \cdot \delta_{ij} = 2\mu\eta_{ij}, \\ &\sigma_{12}\big|_{x=0} = 0. \end{aligned} \tag{4.2}$$

The first of these equations is the static Beltrami equation in the presence of external forces, and the second one is the condition of equality to zero of the elastic stresses at any section of the bent boundary, as the consequence of its equilibrium equation. As in the previous section, it is assumed that the domain boundary in the elastic separates the domains characterised by spontaneous deformation $+\epsilon_0, -\epsilon_0$, and the plane of the nondisplaced domain wall coincides with the $zy$ coordinate plane.

Let us assume that the centre of dilatation is located at the point with coordinates $\mathbf{r} = (x_d, 0, 0)$. Let us find the distribution of the stresses in the system. As in the previous section on the basis of the Fourier expansion (3.6) here

$$\begin{aligned} &\tilde{\sigma}''_{11} - k^2\tilde{\sigma}_{11} + \beta k^2\tilde{\sigma} + \tilde{f}'_1 - ik_y\tilde{f}_2 - ik_z\tilde{f}_3 = 0, \\ &\tilde{\sigma}''_{22} - k^2\tilde{\sigma}_{22} + \beta\left(\tilde{\sigma} \cdot k_z^2 - \tilde{\sigma}''\right) - \tilde{f}'_1 - ik_y\tilde{f}_2 - ik_z\tilde{f}_3 = 0, \\ &\tilde{\sigma}''_{33} - k^2\tilde{\sigma}_{33} + \beta\left(\tilde{\sigma} \cdot k_y^2 - \tilde{\sigma}''\right) - \tilde{f}'_1 - ik_y\tilde{f}_2 + ik_z\tilde{f}_3 = \\ &\qquad\qquad = -4\mu\epsilon_0 ik_y U_{\mathbf{k}} \cdot \delta'(x), \end{aligned} \tag{4.3}$$

and adding up these equations we obtain

$$\tilde{\sigma}'' - k^2\tilde{\sigma} = \frac{\tilde{f}'_1 + ik_y\tilde{f}_2 + ik_z\tilde{f}_3}{(2\beta - 1)} + \frac{4\mu\epsilon_0\delta'(x) ik_y U_{\mathbf{k}}}{(2\beta - 1)}, \tag{4.4}$$

whence

$$\sigma_{k_x} = \frac{\left(\lambda + \frac{2}{3}\mu\right)}{(2\beta - 1)} \cdot \Omega_0 \cdot \exp(-ik_x x_d) + \frac{4\mu\epsilon_0 k_x k_y U_{\mathbf{k}}}{(2\beta-1)\left(k_x^2 + k^2\right)}. \qquad (4.5)$$

The Fourier image $\tilde{\sigma}_{12}$ is determined here by the equation

$$\tilde{\sigma}''_{12} - k^2\tilde{\sigma}_{12} + \beta i k_y \tilde{\sigma}' + 2ik_y \tilde{f}_1 = 2\mu\epsilon_0 k_z^2 \delta(x) U_{\mathbf{k}}, \qquad (4.6)$$

whence taking into account (4.5)

$$\begin{aligned}\sigma_{12}^{k_x} &= -\frac{2\mu\epsilon_0 k_z^2}{\left(k_x^2 + k^2\right)} \cdot U_{\mathbf{k}} - \frac{\beta k_x k_y}{\left(k_x^2 + k^2\right)} \sigma_{k_x} + \frac{2ik_y f_1^{k_x}}{\left(k_x^2 + k^2\right)} = \\ &= -\frac{2\mu\epsilon_0 U_{\mathbf{k}}}{\left(k_x^2 + k^2\right)} \cdot \left[k_z^2 + \frac{2\beta}{(2\beta - 1)} \cdot \frac{k_x^2 k_y^2}{\left(k_x^2 + k^2\right)}\right] + \\ &+ \frac{\left(\lambda + \frac{2}{3}\mu\right)\Omega_0 k_x k_y}{\left(k_x^2 + k^2\right)} \cdot \exp(ik_x x_d) \cdot \left[2 - \frac{\beta}{(2\beta - 1)}\right].\end{aligned} \qquad (4.7)$$

Calculating $\tilde{\sigma}_{12}(x)$ on the basis of (4.7) gives

$$\begin{aligned}\tilde{\sigma}_{12}(x) &= \tilde{\sigma}_{12}^1 + \tilde{\sigma}_{12}^2 + \tilde{\sigma}_{12}^3, \\ \tilde{\sigma}_{12}^1 &= -\mu\epsilon_0 U_{\mathbf{k}} \cdot \exp(-k|x|) \cdot \frac{k_z^2}{k}, \\ \tilde{\sigma}_{12}^2 &= -\mu\epsilon_0 U_{\mathbf{k}} \cdot \frac{\beta}{(2\beta - 1)} \cdot \frac{k_y^2}{k}(1 - k|x|)\exp(-k|x|), \\ \tilde{\sigma}_{12}^3 &= \frac{\mu}{3} \cdot \frac{(3\lambda + 2\mu)}{(\lambda + 2\mu)} \cdot \Omega_0 i k_y \cdot \exp(-k|x - x_d|).\end{aligned} \qquad (4.8)$$

Equating to zero the sum $\tilde{\sigma}_{12}(x)$ at $x$=0, we find the Fourier image of the displacement of the boundary [108]:

$$U_{\mathbf{k}} = \frac{(3\lambda + 2\mu)}{3(\lambda + 2\mu)} \cdot \frac{\Omega_0 i k_y k \cdot \exp(-k|x_d|)}{\epsilon_0\left[k_z^2 + \omega k_y^2\right]}, \quad \omega = \frac{2(\lambda + \mu)}{(\lambda + 2\mu)}. \qquad (4.9)$$

The energy of interaction of the dilatation centre with the domain boundary of the ferroelastic is determined by the trace of the induced part of stresses

$$\mathcal{U}_0 = -\frac{1}{3}\Omega_0 \cdot \sigma_{kk}^{ind}. \tag{4.10}$$

Substituting (4.8) into (4.5) and then into (4.10) shows that the energy of interaction decreases with increasing distance $x_d$ in proportion to $1/x_d^3$. At the same time the maximum energy of interaction, represented by its value at $x_d = \delta$ is equal to

$$\mathcal{U}_0 = \frac{1}{18\pi^2}\frac{(3\lambda+2\mu)^2}{(\lambda+2\mu)^2}\cdot\mu\Omega_0^2\cdot\frac{1}{\delta^3}. \tag{4.11}$$

(in calculations of $\mathcal{U}_0$ it was assumed that $\omega \simeq 1$ in order to simplify calculations).

Using in calculations the values of $\mu \sim 10^{10}$ CGSE units, $d \sim 10^{-7}$ cm, where $a^3$ is the atomic volume and the value of $a$ is equal to approximately half the size of the elementary cell, whose typical value is $10^{-7}$ cm, we obtain $\mathcal{U}_0 \sim 0.02$ eV.

The distribution of displacements of the boundary, interacting with the defect at $\omega \simeq 1$, is described by the function (Fig.4.5)

$$U(z,y) = -\frac{1}{3\pi}\frac{(3\lambda+2\mu)}{(\lambda+2\mu)}\cdot\frac{\Omega_0}{\epsilon_0}\cdot\frac{y}{\left(y^2+z^2+x_d^2\right)^{3/2}}. \tag{4.12}$$

As we can see from Fig. 4.5, the displacement of the wall is symmetric in direction normal to spontaneous shift and asymmetric along this direction.

Concluding the discussion of the interaction of point defects with domain boundaries, let us briefly consider here the case of lattice disruptions similar to them – the so-called non-ferroelectric inclusions. The interaction of these defects with the domain walls is associated with the finiteness of their dimensions and is determined by at least two reasons.

On the one hand, the occurrence of the non-ferroelectric inclusion directly on the boundary decreases its area (Fig.4.6). Since the energy of the boundary has the positive sign, the decrease of the boundary surface means the attraction of the boundary to the defect. The mentioned interaction is evidently of a short-range nature because it occurs only if the defect falls directly on the boundary, and its maximum value is [109, 110]

$$\mathcal{U}_{0max} = \gamma\pi R^2. \tag{4.13}$$

Here $\gamma$ is the surface density of energy of the domain wall, and $R$ is the radius of the inclusion.

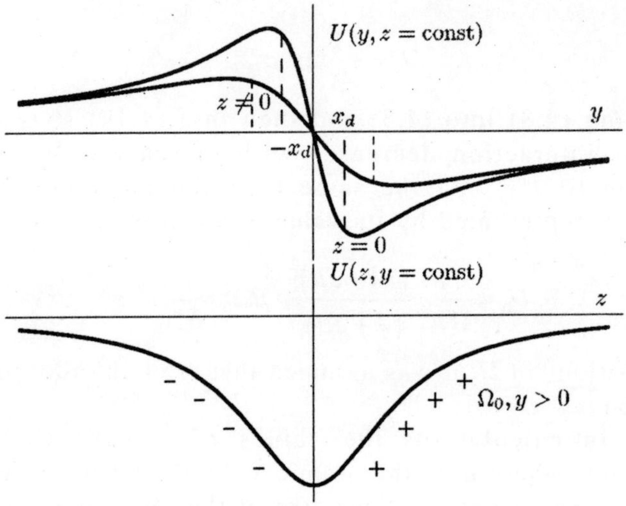


**Fig.4.5.** Distribution of displacements of the domain wall of the ferroelastic interacting with the dilation centre.

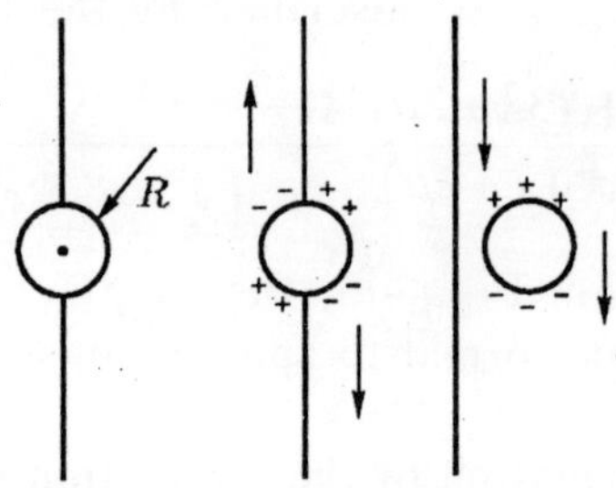


**Fig.4.6.** Interaction of the domain boundary with a non-ferroelectric inclusion as a result of (a) – a decrease of the area of the boundary, (b) – as a result of a decrease of the energy of the depolarizing field of the inclusion.

On the other hand, the occurrence of the defect on the boundary decreases the energy of the bound charges of spontaneous polarization, formed on the surface of the inclusion (Fig.4.6), as in the case of the partitioning of crystals into domains. This interaction is also of a short-range nature with the maximum energy of interaction of the defect with the boundary equal to [111]:

$$\mathcal{U}_{0max} = \frac{\pi P_0^2}{\sqrt{\varepsilon_a \varepsilon_c}} \cdot R^3. \tag{4.14}$$

## 4.5. Interaction of the ferroelastic domain boundary with a dislocation parallel to the plane of the boundary

In addition to point defects the crystals also contain linear defects – dislocations. An edge dislocation is a perturbation caused by an extra half plane inserted in the lattice. A screw dislocation is the result of 'sectioning' the lattice along the half plane with subsequence shift of the sectioned parts parallel to the edge of the section. For usual, i.e. non-twinning dislocations, the Burger's vector **b**, equal to the increment to the vector of elastic displacements of the medium when traversing the closed contour around the dislocation line coincides with one of the lattice spacings. For the edge dislocation, the unique vector of the tangent to dislocation line $\boldsymbol{\tau}\perp\mathbf{b}$, and for the screw dislocation $\boldsymbol{\tau}\|\mathbf{b}$.

As in the case of interaction with the dilatation centre, during interaction of a domain boundary in ferroelastic with a dislocation certain boundary bending can remove part of the stresses generated in the crystal by the dislocation. This efficiently indicates their mutual attraction.

When describing the interaction of a specific type dislocation, with a domain boundary in ferroelastic it is necessary to specify their position relationship. Let us consider dislocations parallel and normal to the plane of the boundary. Let as in the previous section the initial position of the boundary coincide with the $zy$-plane and the direction of spontaneous shear with the $y$ axis. The distribution of elastic stresses and the profile of the bent boundary together with it are determined here by set of equations (4.2) in which the components of the volume density of forces $f_i$ can be assumed to be equal to zero

$$\begin{gathered}\sigma_{ij,kk}+\frac{m}{m+1}\left(\sigma_{kk,ij}-\sigma_{kk,ll}\cdot\delta_{ij}\right)=2\mu\eta_{ij},\\ \sigma_{12}\big|_{x=0}=0,\end{gathered}\tag{5.1}$$

and the components of the tensor of incompatibility are determined not only by the density of the twinning dislocations (3.4)

$$\begin{aligned}\alpha_{22}&=2\epsilon_0\cdot\frac{\partial U}{\partial z}\delta(x),\\ \alpha_{32}&=-2\epsilon_0\cdot\frac{\partial U}{\partial y}\delta(x),\end{aligned}\tag{5.2}$$

formed at the boundary bending, but also by the densities of the initial dislocations, interacting with the boundary.

In order to write down the components of the tensor $\alpha^0_{in}$ we rewrite equation (2.3) for an individual dislocation in a trifle different way. Since tensor $e_{ikl}$ is antisymmetric in respect of the indices $k, l$ and tensor $\partial u_{ln}/\partial x_k = \partial^2 u_n/\partial x_k \partial x_l$ – is symmetric in respect of the same indices, then integrand in (2.2) is in fact equal to zero everywhere with the exception of the point of intersection of the dislocation line with the surface on which integration is carried out. Taking into account this and initial definition (2.1) for the individual dislocation, the subintegrand in (2.2) can be presented in the form

$$-\alpha^0_{in} = e_{ikl} \cdot \frac{\partial u_{ln}}{\partial x_k} = -\tau_i b_n \delta(\xi), \tag{5.3}$$

where $\xi$ is the two-dimensional radius-vector counted from the axis of the dislocation in the plane normal to the vector $\boldsymbol{\tau}$. It may easily be seen that in accordance with the definition of the dislocation, integration of (5.3) over $d\Sigma_i$ gives the component of Burger's vector $b_n$.

Let us consider first of all the interaction with the boundary of dislocations whose line is parallel to the boundary plane. Let us start with the edge dislocation. It can be easily seen that if the axis of the edge dislocation is parallel to the direction of spontaneous shear, then it does not interact with the boundary. In fact for such a dislocation, the components of the stress tensor that differ from zero are $\sigma_{11}$ and $\sigma_{13}$, and component $\sigma_{12}$ is missing. In this case, in accordance with (5.1) there is no pressure on the boundary, its displacement is equal to zero everywhere, and consequently no stresses occur in the location of dislocation.

For an edge dislocation normal to spontaneous shear, i.e., in this case, parallel to axis $z$ and the line of the dislocation described by the radius vector $\boldsymbol{\rho}_d=(x=x_d,\ y=0)$ the component of tensor $\alpha^0_{in}$ that differs from zero is the following one

$$\alpha^0_{32} = b \cdot \delta(x - x_d) \cdot \delta(y), \tag{5.4}$$

to which corresponds the only non-zero component of the incompatibility tensor

$$\eta_{33} = \frac{\partial \alpha^0_{32}}{\partial x} = b \cdot \delta'(x - x_d) \cdot \delta(y). \tag{5.5}$$

On the basis of Beltrami's equation in (5.1) taking into account the homogeneity of the discussed fields and displacements of the boundary along the dislocation axis as well as ratio (5.5), the expression for the Fourier image of the trace of the part of the matrix of elastic stresses $\sigma_{ij}$, which is formed by the dislocation that interacts with the boundary, has the form

$$\tilde{\sigma}_{k_x} = -\frac{ik_x \cdot 2\mu b \exp(ik_x x_d)}{(1-2\beta)\left(k_x^2+k_y^2\right)}. \tag{5.6}$$

Similarly, the Fourier image

$$\tilde{\sigma}_{12}^{k_x} = \frac{-\beta \cdot k_x k_y}{\left(k_x^2+k_y^2\right)} \cdot \tilde{\sigma}_{k_x}. \tag{5.7}$$

Substitution of (5.6) into (5.7) gives

$$\tilde{\sigma}_{12}^{k_x} = \frac{2\mu b \cdot k_x^2 ik_y \cdot \exp(-ik_x x_d)}{(1-2\beta)\left(k_x^2+k_y^2\right)^2}, \tag{5.8}$$

whence taking into account integration of (5.8) over $k_x$

$$\tilde{\sigma}_{12}(x=0) = \frac{i\mu b}{2(1-2\beta)} \operatorname{sign} k_y \left(1 - x_\alpha k_y\right) \cdot \exp(-k_y x_d). \tag{5.9}$$

Adding up equation (5.9) with the field of elastic stresses (3.16) induced by the bent boundary, and taking into account the homogeneity along the $z$ axis (in this case in (3.16) $k_z=0$), from the equation of the boundary equilibrium in (5.1) we obtain the Fourier image of displacement of the ferroelastic domain boundary bent as a result of interaction with the edge dislocation [112, 113]:

$$U_{\mathbf{k}} = \frac{-ib}{2\beta\epsilon_0} \cdot \frac{1}{k_y} \cdot \left(1 - x_d k_y\right) \cdot \exp(-k_y x_d). \tag{5.10}$$

Consequently, the coordinate dependence of the wall displacement is

$$U(y) = \frac{b}{2\pi\beta\epsilon_0} \cdot \left[\operatorname{arctg}\frac{y}{x_d} - \frac{x_d y}{x_d^2 + y^2}\right]. \tag{5.11}$$

The dependence expressed by (5.11) is shown in Fig.4.7 that indicates that the displacement of the free wall represents a kink where the twinning dislocations of the unlike sign as compared to the initial dislocation are located.

The width of the kink is equal to $2x_d$, i.e. increases with increase of the distance of the dislocation from the boundary. At $x_d \to 0$ the kink at the boundary transforms to a step with the height of $b/2\epsilon_0\beta$.

When describing the force and energy characteristics of interaction of the dislocation with the boundary it is convenient to calculate the force acting on the dislocation from the direction of the stresses induced by the dislocation.

The equation for the components of this force is described by the so-called Peach–Koehler force [100]

$$f_i = e_{ikl}\tau_k\sigma_{lm}^{ind}b_m. \tag{5.12}$$

The direct calculations and analysis of the symmetry of this problem show that the only component that differs from zero in this case is:

$$f_1 = -\sigma_{22}^{ind}\left(y=0, x=x_d\right)\cdot b. \tag{5.13}$$

To determine this component, it is first of all necessary to find the component of the stress tensor $\sigma_{22}$ induced by bending of the boundary. According to (5.1), the equation for the two-dimensional Fourier image $\tilde{\sigma}_{22}$ has the form:

$$\tilde{\sigma}_{22}'' - k_y^2\tilde{\sigma}_{22} - \beta\tilde{\sigma}'' = 0, \tag{5.14}$$

where the primes, as previously, indicate differentiation in respect of $x$. Hence, taking into account (3.10) and the specific form of the Fourier image $U_{\mathbf{k}}$ (5.10)

$$\tilde{\sigma}_{22}^{k_x} = \frac{\beta\cdot k_x^2}{\left(k_x^2+k_y^2\right)}\cdot\tilde{\sigma}_{k_x} = \frac{4\mu\epsilon_0\beta\cdot k_x^3k_y}{(2\beta-1)\left(k_x^2+k_y^2\right)^2}\cdot U_{\mathbf{k}} =$$

$$= \frac{-2\mu i k_x^3 b}{(2\beta-1)\left(k_x^2+k_y^2\right)^2}\cdot\left(1-x_dk_y\right)\cdot\exp(-x_dk_y). \tag{5.15}$$

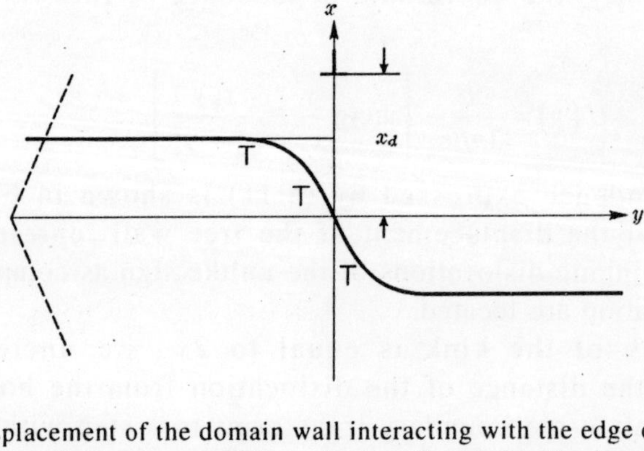


**Fig.4.7.** Displacement of the domain wall interacting with the edge dislocation. Thin line shows the displacement of the wall at $x_d$=0 and also the image dislocation.

The Peach–Koehler force taking into account (5.15) is:

$$f \equiv f_1 = -b\int_{-\infty}^{\infty}\int_{-\infty}^{\infty}\tilde{\sigma}_{22}^{k_x}\cdot\exp(ik_x x_d)\frac{dk_x dk_y}{(2\pi)^2} = \frac{-\mu b^2}{2\pi(2\beta-1)}\frac{1}{2x_d}. \quad (5.16)$$

It can be easily seen that the interaction (5.15), as in the previous problems where we examined the free boundary and a 'pure' material, i.e. a ferroelectric or ferroelastic, is the interaction with the image. In this case the image also is the edge dislocation (Fig. 4.7), which as any image has the unlike sign as compared to the original and, therefore, their interaction at any sign of the initial dislocation represents mutual attraction. The characteristic law of the decrease of the function $f(x_d)$ with the distance (Fig.4.8) is the law $f(x_d)\sim 1/x_d$. At $x_d\to 0$ the value of the Peach–Koehler force increases to the maximum, which is restricted by the value of $f$ at higher of the values of $\delta$ or $U_{max}$, where $\delta$ is the half width of the domain wall, and $U_{max}$ is the maximum displacement of the boundary, which restricts the possibility of application of the linear approximation for this problem. At conventional values of $\epsilon_0\sim 10^{-2}$, $\beta\sim 1$, the value $U_{max}\sim 10^2 a>\delta$ and, therefore, exactly this value determines the maximum force of interaction, which is equal to

$$|f_{max}| = \frac{\mu b\epsilon_0\beta}{\pi(2\beta-1)}. \quad (5.17)$$

The energy of interaction of the boundary with the dislocation, related to its unit length, can be calculated as the work of carrying over of the dislocation from the point with the coordinate $x_d=U_{max}$ to infinity. This gives the following equation

$$\mathcal{U}_{0\tau} = -\frac{\mu b^2}{4\pi(2\beta-1)}\cdot\ln\frac{L}{U_{max}}, \quad (5.18)$$

where $L$ is the characteristic size of the crystal.

Let us consider the interaction of a screw dislocation, the line of which is parallel to the initial position of the domain boundary, with the domain wall of a ferroelastic. Among screw dislocations orientated in this direction the dislocations with vector $\boldsymbol{\tau}$, normal to the direction of spontaneous shear, do not generate stresses $\sigma_{12}$, that are active in the displacement of the boundary and, therefore, they do not interact with it. On the contrary, the dislocations whose vector $\boldsymbol{\tau}$ is parallel to spontaneous shear, generate stresses and interact with the domain boundary. To describe this interaction, let us find an expression for the stresses created by the dislocation

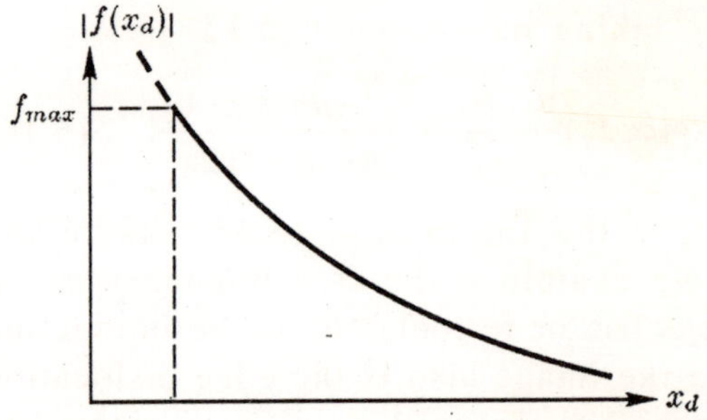


**Fig.4.8.** Dependence of the force of interaction of the dislocation with the boundary on the distance between them.

under consideration in the location of the domain boundary. The component of the tensor of dislocation density $\alpha_{in}^{0}$, which differs from zero is equal in this case to

$$\alpha_{22}^{0}=b\cdot\delta\left(x-x_{d}\right)\cdot\delta\left(z\right). \tag{5.19}$$

The non-zero components of the incompatibility tensor correspond to this component

$$\eta_{12}=-\frac{1}{2}\frac{\partial\alpha_{22}^{0}}{\partial z}=-\frac{b}{2}\cdot\delta\left(x-x_{d}\right)\cdot\delta'\left(z\right),$$
$$\eta_{23}=\frac{1}{2}\frac{\partial\alpha_{22}^{0}}{\partial x}=\frac{b}{2}\cdot\delta'\left(x-x_{d}\right)\cdot\delta\left(z\right). \tag{5.20}$$

Taking into account (5.20) from the Beltrami equation $\tilde{\sigma}=0$ and the Fourier image

$$\sigma_{12}^{k_x}=\frac{\mu bik_z\cdot\exp(-ik_x x_d)}{\left(k_x^2+k_z^2\right)}. \tag{5.21}$$

Hence,

$$\tilde{\sigma}_{12}\left(x=0\right)=\frac{\mu bik_z}{2k_z}\cdot\exp(-k_z x_d). \tag{5.22}$$

The Fourier image of the stresses, induced by the boundary bending taking into account the homogeneity of all values along the dislocation axis, i.e., the $y$ axis from (3.16) is equal to

$$\tilde{\sigma}_{12}^{ind}\left(x=0\right)=-\mu\,\epsilon_0 U_{\mathbf{k}}k_z\,\mathrm{sign}\,k_z. \tag{5.23}$$

Equating of sums (5.22) and (5.23) to zero yields the equation for determining the Fourier image of displacement of the boundary, whence

$$U_{\mathbf{k}}=\frac{i\cdot b}{2\epsilon_0 k_z}\cdot\exp(-k_z x_d). \tag{5.24}$$

The calculation of the coordinate dependence of the boundary displacement on the basis of (5.24), shows that the resultant dependence

$$U(z) = -\frac{b}{2\pi\epsilon_0}\cdot \operatorname{arctg}\frac{z}{x_d} \tag{5.25}$$

is qualitatively similar to the boundary displacement in the field of the edge dislocation (Fig.4.7).

The force of interaction of the domain boundary with the screw dislocation of the given orientation has got one component different from zero

$$f_1 = b\cdot\sigma_{23}^{ind}(x = x_d, z = 0). \tag{5.26}$$

The Beltrami equation for the component $\sigma_{23}$ has the form

$$\tilde{\sigma}_{23}'' - k_z^2\tilde{\sigma}_{23} = 2\mu\cdot\eta_{23}, \tag{5.27}$$

where the induced part of this component $\sigma_{23}^{ind}$ is determined by the part of the component $\eta_{23}$ related to the boundary bending (3.5) whence

$$\sigma_{12}^{k_x ind} = \frac{2\mu\varepsilon_0 k_x k_z}{(k_x^2 + k_z^2)}\cdot U_k = \frac{\mu\cdot bik_x}{(k_x^2 + k_z^2)}\cdot \exp(-k_z x_d). \tag{5.28}$$

Taking into account (5.28), the component of the force

$$f_1 = -\frac{\mu b^2}{4\pi}\cdot\frac{1}{x_d} \tag{5.29}$$

and the linear density of the energy of interaction of the boundary with the dislocation here is equal to

$$\mathcal{U}_{0\tau} = \frac{\mu b^2}{4\pi}\cdot \ln\frac{L}{U_{max}}, \quad U_{max} = b/4\epsilon_0 \cdot \tag{5.30}$$

## 4.6. Interaction of the domain boundary of a ferroelastic with the dislocation perpendicular to the boundary plane

In addition to the dislocations, parallel to the plane of the domain wall, the crystal evidently also contains dislocations inclined with regard to the wall. Let us discuss the interaction of a domain wall with a dislocation intersecting it for the most symmetric situation when the dislocation is perpendicular to the boundary.

Let us assume that as before the plane of a non-perturbed boundary coincides with the *zy* coordinate plane, and the direction of spontaneous shear with the axis *y*. For a screw dislocation, perpendicular to the boundary, the component of the tensor of dislocation density $\alpha_{in}^{0}$ different from zero is equal to

$$\alpha_{11}^{0}=b\delta(y)\delta(z), \tag{6.1}$$

and the non-zero components of the incompatibility tensor corresponding to this component are:

$$\begin{aligned}\eta_{12}&=\frac{1}{2}\frac{\partial\alpha_{11}^{0}}{\partial z}=\frac{b}{2}\cdot\delta(y)\cdot\delta'(z),\\ \eta_{13}&=\frac{1}{2}\frac{\partial\alpha_{11}^{0}}{\partial y}=-\frac{b}{2}\cdot\delta'(y)\cdot\delta(z).\end{aligned} \tag{6.2}$$

Taking into account equation (6.2), from the Beltrami equation, the Fourier image of the part of the shear stress $\sigma_{12}$ generated by the initial dislocation, is

$$\tilde{\sigma}_{12}=\frac{\mu bik_z}{\left(k_y^2+k_z^2\right)}. \tag{6.3}$$

Adding to (6.3) the Fourier image of the stresses $\sigma_{12}$ induced by the boundary bending and written in the general form (3.16)

$$\tilde{\sigma}_{12}(x=0)=-\frac{\mu\epsilon_0}{k}\cdot U_{\mathbf{k}}\left(k_x^2+\omega k_y^2\right),\quad \omega=\frac{2(\lambda+\mu)}{\lambda+2\mu} \tag{6.4}$$

and equating the resultant sum to zero, we obtain the Fourier image of the boundary displacement

$$U_{\mathbf{k}}=\frac{b}{\epsilon_0}\cdot\frac{ik_z}{\sqrt{k_y^2+k_z^2}.\left(k_x^2+\omega k_y^2\right)}. \tag{6.5}$$

From this, the displacement of the boundary is

$$U(y,z)=\frac{b}{2\pi\epsilon_0}\cdot\sqrt{\frac{\lambda+2\mu}{\lambda}}\cdot\operatorname{arctg}\left(\sqrt{\frac{\lambda}{\lambda+2\mu}}\cdot\frac{z}{\sqrt{z^2+y^2}}\right). \tag{6.6}$$

As can be seen from the symmetry of the problem for an infinite dislocation in this case there is no specific position with respect to the boundary, and consequently the Peach–Koehler force in this case is equal to zero. Calculation of this force for the dislocation of finite dimensions is difficult and, therefore, to determine the energy of interaction of the boundary with the dislocation perpendicular to it with at least one of the dimensions - either of boundary or of dislocation – being finite, one should use a different procedure. The value of this dimension can be conveniently found as the difference between the elastic energy of the system consisting of the dislocation intersecting the boundary, and the elastic energy of a separate dislocation. In order to determine it, let us find the distribution of the elastic stresses in the system under consideration. Taking into account (3.5), (6.5) and (6.2), the Fourier images of the total components of the stresses here are equal to [114]

$$\tilde{\sigma}_{11} = \frac{4\mu bik_x k_y k_z \cdot k \cdot \omega}{\left(k_x^2 + k^2\right)^2 \cdot \left(k_z^2 + \omega k_y^2\right)},$$

$$\tilde{\sigma}_{22} = \frac{4\mu bik_x k_y k_z \left(k_z^2 + k_x^2\right) \cdot \omega}{k\left(k_x^2 + k^2\right)^2 \cdot \left(k_z^2 + \omega k_y^2\right)},$$

$$\tilde{\sigma}_{33} = \frac{4\mu bik_x k_y k_z}{k\left(k_z^2 + \omega k_y^2\right)} \cdot \left[\frac{\left(k_y^2 + k_x^2\right) \cdot \omega}{\left(k_x^2 + k^2\right)} - 1\right],$$

$$\tilde{\sigma}_{12} = \frac{2\mu bik_z}{k\left(k_x^2 + k^2\right)\left(k_z^2 + \omega k_y^2\right)} \cdot \left[-k_z^2 - \frac{2k_y^2 k_x^2 \cdot \omega}{\left(k_x^2 + k^2\right)}\right] + \frac{\mu bik_z \delta(k_x)}{\left(k_x^2 + k^2\right)},$$

$$\tilde{\sigma}_{13} = \frac{2\mu bik_y k_z^2}{k\left(k_x^2 + k^2\right)\left(k_z^2 + \omega k_y^2\right)} \cdot \left[1 - \frac{2k_x^2 \cdot \omega}{\left(k_x^2 + k^2\right)}\right] + \frac{\mu bik_z \delta(k_x)}{\left(k_x^2 + k^2\right)}, \qquad (6.7)$$

The elastic energy of the system is

$$W = \frac{1}{4\mu} \iint \Big[ 2\left(\sigma_{12}^2 + \sigma_{13}^2 + \sigma_{23}^2\right) + \sigma_{11}^2 + \sigma_{22}^2 + \sigma_{33}^2 - \\ - \frac{\lambda}{(3\lambda + 2\mu)} \cdot \sigma_{pp}^2 \Big] \frac{d\mathbf{k} dk_x}{(2\pi)^3}. \qquad (6.8)$$

The range of integration in (6.8) in the plane $(k_y, k_z)$ is a circular ring with the internal radius $1/L_1$, where $L_1$ is the characteristic size of the domain boundary and with the external radius of $4\epsilon_0/b$. After integration in (6.8) we find

$$W = \frac{\mu b^2}{4\pi} \cdot L \cdot \ln\frac{L_1}{b} - \frac{L_1 - (b/4\epsilon_0)}{2\pi} \cdot \frac{\mu b^2}{1 + \sqrt{\lambda/(\lambda + 2\mu)}}. \qquad (6.9)$$

$$\tilde{\sigma}_{23} = \frac{2\mu bik_y k_z^2}{k\left(k_x^2 + k^2\right)\left(k_z^2 + \omega k_y^2\right)} \cdot \left[1 - \frac{2k_y^2 \cdot \omega}{\left(k_x^2 + k^2\right)}\right].$$

The first term here is the intrinsic elastic energy of the screw dislocation, and the second one describes the decrease of the elastic energy of the system as a result of the domain wall bending in the elastic field of the dislocation, $L$ is the dislocation length.

Taking into account that $L_1 \gg b/4\epsilon_0$, and dividing the second term in (6.9) by the dislocation length, we obtain in this case the following equation for the mean linear density of the energy of interaction of the boundary with the dislocation

$$\mathcal{U}_{0\tau} = \frac{L_1}{L} \cdot \frac{\mu b^2}{2\pi \cdot \left[1 + \sqrt{\lambda/(\lambda + 2\mu)}\right]}. \tag{6.10}$$

For the edge dislocation, intersecting the domain boundary along the perpendicular with Burger's vector **b** parallel to either $y$ or $z$ axis, there is no interaction with the boundary.

## Chapter 5

# Structure of domain boundaries in real ferroactive materials

## 5.1. ORIENTATION INSTABILITY OF THE INCLINED DOMAIN BOUNDARIES IN FERROELECTRICS. FORMATION OF ZIG-ZAG DOMAIN WALLS

Like formations with a positive energy, the domain boundaries in ferroelectrics and ferroelastics tend to minimize their surface. And due to that in the absence of other competing factors, they usually become flat. The permissible boundaries of the flat type, and also the dependence of the energy of such boundaries on the orientation have been considered in [115–121]. However, there is large number of situations in which the broken or deformed profile is more advantageous for the domain boundaries [122–128]. In this case, the evident loss due to the increase of the total surface of the wall is compensated by the energy gained in some other way. For an inclined domain wall it is the gain in the electrostatic energy of charges on the wall due to such an orientation in which the density of charges on the wall is lower. In crystals with the phase transitions of the first order, this gain is achieved as a result of transition of a part of the crystal volume in the field of charges on the bent boundary from the metastable phase to a phase with more advantageous energy parameters. In crystals with defects, the mentioned gain is produced by the reduction of the energy of the system – domain boundary plus defects as the result of capture of certain sections of the boundary by defects. All the above situations, and also the situations with the thermal distortion of the boundary profile, are considered in this chapter.

Let us in the first place discuss the situation with an inclined domain boundary in a uniaxial ferroelectric. Let the polarization

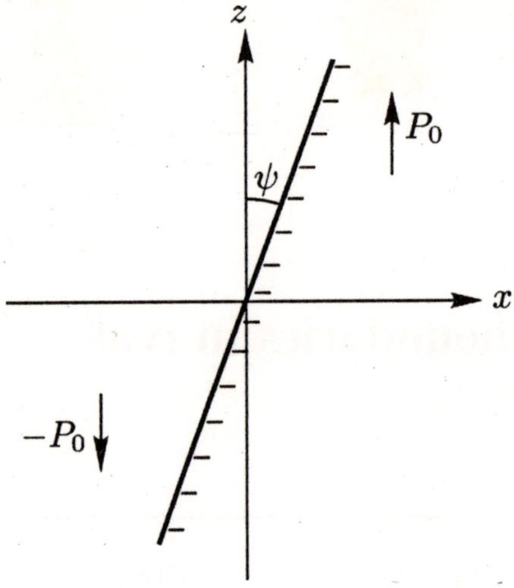


**Fig. 5.1.** Inclined 180° domain boundary in a uniaxial ferroelectric.

vector be directed along the $z$ axis. The plane of the domain boundary is initially located in the crystallographic plane $zy$. To be more specific it is assumed that in the right domain ($x>0$) the polarization vector **P** is oriented in the positive direction of the axis z and in the left domain ($x<0$) in the negative direction. Let us rotate the plane of the domain boundary through the angle $\psi$ around the $y$ axis (Fig.5.1).

In this case a charge is formed on the domain boundary, and the electric field

$$\mathbf{E} = -4\pi(\mathbf{P},\mathbf{n})\mathbf{n} \tag{1.1}$$

appears in the bulk of the crystals. Here **n** is the normal to the plane of the boundary between the domains, directed inside the domain, in which the electric field is specified (1.1). In particular, for the right domain $\mathbf{n} = (\cos\psi,\ \sin\psi)$. Taking into account the presence of the field, the volume density of the thermodynamic potential is

$$\Phi = \Phi_0(\mathbf{P}) + \frac{E^2}{8\pi}, \tag{1.2}$$

where for the uniaxial ferroelectric with the phase transition of the second order (the general considerations also apply to ferroelectrics with the phase transition of the first order), thermodynamic potential $\Phi_0(\mathbf{P})$, linked with the short-range interatomic forces, is

$$\Phi_0(\mathbf{P}) = \frac{\alpha_x}{2}P_x^2 - \frac{\alpha_z}{2}P_z^2 + \frac{\beta}{4}P_z^4. \tag{1.3}$$

Substitution of (1.1) into (1.2) gives

$$\Phi = \frac{\alpha_x}{2}P_x^2 - \frac{\alpha_z}{2}P_z^2 + \frac{\beta}{4}P_z^4 + 2\pi\left(P_x\cos\psi - P_z\sin\psi\right)^2. \qquad (1.4)$$

Variation of $\Phi$ in respect of the components of the polarization vector leads to the equations

$$\frac{\partial\Phi}{\partial P_x} = \alpha_x P_x + 4\pi\cos\psi\left(P_x\cos\psi - P_z\sin\psi\right) = 0, \qquad (1.5)$$

$$\frac{\partial\Phi}{\partial P_z} = -\alpha_z P_z + \beta P_z^3 - 4\pi\sin\psi\left(P_x\cos\psi - P_z\sin\psi\right) = 0. \qquad (1.6)$$

Hence

$$P_x = \frac{4\pi\sin\psi\cos\psi}{\alpha_x + 4\pi\cos^2\psi}P_z, \qquad (1.7)$$

$$P_{z0}^2 = \frac{\tilde{\alpha}_z}{\beta}, \quad \tilde{\alpha}_z = \alpha_z - \frac{\alpha_x\cdot 4\pi\sin^2\psi}{\alpha_x + 4\pi\cos^2\psi}. \qquad (1.8)$$

For the case of $\alpha_x$, $\alpha_z$, $\psi << 1$ that is important in practice

$$P_x = P_z\cdot\sin\psi, \quad \tilde{\alpha}_z = \alpha_z - \alpha_x\sin^2\psi. \qquad (1.9)$$

The formulas (1.8) and (1.9) show that at a relatively large angle of inclination of the boundary $\psi_0 > \sqrt{\alpha_z}/\alpha_x$ the paraphase becomes advantageous from the thermodynamic viewpoint. Thus, one of the possible channels of decrease of the energy of the inclined domain boundary is the volume channel associated with the instability of polarization of the bulk of the crystal in the field of the bound charge, generated by the inclined domain boundary.

It should be mentioned that this process is restricted in the best case by the layer with the thickness of $2d$ in the vicinity of the inclined domain wall, where $d$ is the width of the domains into the system of which the material is divided (Fig. 5.2) in order to reduce the energy of the depol-

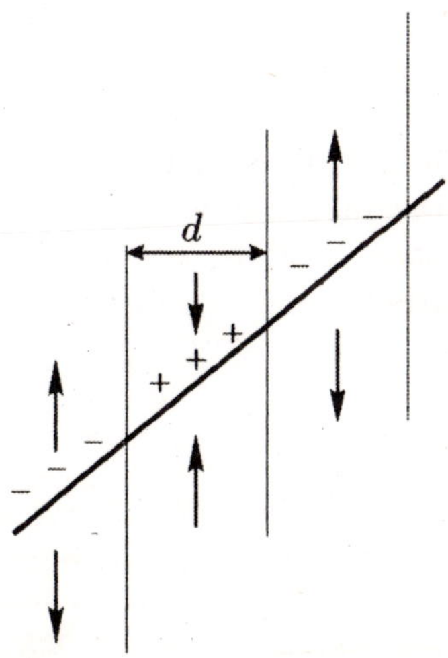


**Fig.5.2.** Formation of the substructure of the domains in the vicinity of the inclined domain boundary.

arizing field of charges on the boundary. In the given case it is similar to the situation in which the formation of the domain structure in the conventional case reduces the energy of charges of spontaneous polarization on the surface of the ferroelectric material.

Let us assess the minimum period of the new domain structure $d$ from the condition of equality of the density of the thermodynamic potential (1.2) for the material located in the field of charges on the inclined boundary and the similar characteristic for the ferroelectric divided into a system of parallel domains:

$$\Phi_0(\mathbf{P})+\frac{E^2}{8\pi}=\Phi_0(P_0)+\frac{\gamma}{\sin\psi\cdot d}. \tag{1.10}$$

Here $\gamma$ is the surface density of the energy of the symmetric boundary. The appearance of $\sin\psi$ in the denominator of the right hand part of (1.10) describes the effect of elongation of the inclined domain boundary in comparison with the straight domain structure.

At $\alpha_x$, $\alpha_z<1$ taking into account (1.2)

$$\Delta\Phi=\Phi_0(\mathbf{P})-\Phi_0(\mathbf{P}_0)+\frac{E^2}{8\pi}=\frac{\alpha_z\alpha_x\sin^2\psi}{2\beta}, \tag{1.11}$$

and

$$d=\frac{\gamma}{\sin\psi\Delta\Phi}=\frac{\beta\cdot\gamma}{\alpha_z\alpha_x\sin^3\psi}. \tag{1.12}$$

Formula (1.12) shows that the boundary of the infinite dimensions as the idealized system is thermodynamically non-equilibrium for any small angles of rotation $\psi$. However, since the real domain structure is restricted in its size by the dimensions of the crystal $L$, the inclined domain boundaries turn out to be resistant to the formation of the substructure of the domains for small angles of rotation

$$\psi<\frac{1}{L^{1/3}}\sqrt[3]{\frac{\beta\gamma}{\alpha_z\alpha_x}}\simeq\psi_0^{2/3}\sqrt[3]{\frac{\delta}{L}}, \tag{1.13}$$

where $\delta$ is the width of the domain wall, $\psi_0=\sqrt{\alpha_z/\alpha_x}$.

Another possible channel of the loss of stability by the flat inclined domain wall is connected with the change of the geometry of the boundary itself when it becomes zig-zag-shaped (Fig,5.3).

At that the balance is achieved between the decrease of the electrostatic energy of the boundary due to of the increase of the area of the wall (and consequently the region of charge localization)

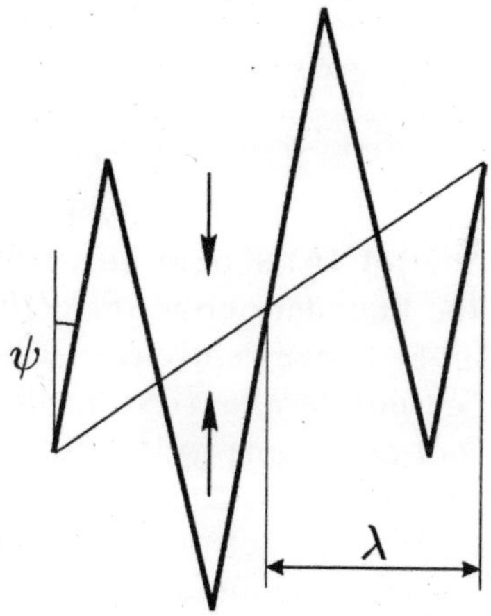


**Fig.5.3.** Zig-zag domain boundary. The initial inclined wall is shown by the thin inclined line.

and the simultaneous increase of its surface energy.

Let us determine the condition of the loss of stability by the flat domain boundary due to the bending of its surface as it was done in [131]. For this purpose we write the thermodynamic potential of a bi-domain ferroelectric crystal with the bent domain boundary:

$$\Phi = \int \left\{ \Phi_0(\mathbf{P}) + \frac{E_x^2 + E_z^2}{8\pi} \right\} dx\, dz + \gamma \int \sqrt{1 + \frac{dz(x)}{dx}} dx. \qquad (1.14)$$

Here the first term describes the volume energy of the ferroelectric and the second term the surface energy of the domain wall whose form is represented by the functions $z = z(x)$. For the purpose of investigation of the loss of stability of the shape of the flat boundary we assume that $z(x) = x\,\mathrm{ctg}\,\psi + U(x)$, where $U(x)$ is the small displacement of sections of the boundary from the average position.

Let us expand equation (1.3) for $\Phi_0(\mathbf{P})$ into a series in the vicinity of $P_{30}^2 = \tilde{\alpha}_z / \beta$ and restrict ourselves to the quadratic term

$$\left( 2\alpha_z - \frac{12\pi\alpha_x \cdot \sin^2\psi}{\alpha_x + 4\pi\cos^2\psi} \right) \cdot \frac{(P_3 - P_{30})^2}{2} + \frac{\alpha_x P_x^2}{2}. \qquad (1.15)$$

Taking into account that $\partial\Phi/\partial P_i = E_i$ equation (1.14) is transformed to the form

$$\Phi = \int \frac{\varepsilon_x E_x^2 + \varepsilon_z E_z^2}{8\pi} dx\, dz + \gamma \int \sqrt{1 + \left( \frac{dz(x)}{dx} \right)^2} dx, \qquad (1.16)$$

where in accordance with (1.15)

$$\varepsilon_x = 1 + \frac{4\pi}{\alpha_x}, \quad \varepsilon_z = 1 + \frac{4\pi}{2\alpha_z - \dfrac{12\pi\alpha_x \cdot \sin^2\psi}{\alpha_x + 4\pi\cos^2\psi}}. \tag{1.17}$$

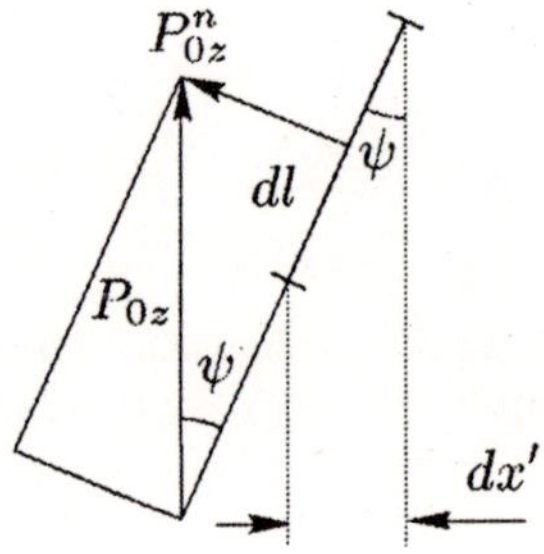


**Fig.5.4.** Calculations of the linear density of the charge on the inclined boundary.

Finally, for the analysis the functional (1.16) should be written using the boundary coordinate $z(x)$. This can be conveniently carried out if we calculate the first term in (1.16) not as the energy of the field in the dielectric medium but as the energy of interaction of bound charges on the boundary. Taking into account (Fig.5.4), that the linear density of the bound charge at the section of the boundary with the length $dl$ is equal to

$$P^n_{0z} \cdot dl = P_{0z} \cdot \sin\psi \cdot \frac{dx'}{\sin\psi} = P_{0z} \cdot dx', \tag{1.18}$$

the electrostatic potential of these charges in the point with the coordinates $(x,z)$ taking into account the anisotropy of the dielectric properties of the ferroelectric can be written in the form:

$$\varphi = -\frac{2P_{0z}}{\sqrt{\varepsilon_x \varepsilon_z}} \cdot \int \ln\left[\frac{(x-x')^2}{\varepsilon_x} + \frac{(z - z(x'))^2}{\varepsilon_z}\right] dx'. \tag{1.19}$$

Then, taking into account (1.19), equation for the functional (1.16) can be presented in the form

$$\Phi = -\frac{2P_{0z}}{\sqrt{\varepsilon_x \varepsilon_z}} \cdot \int\int \ln\left[\frac{(x-x')^2}{\varepsilon_x} + \frac{(z(x) - z(x'))^2}{\varepsilon_z}\right] dx\,dx' + \gamma\sqrt{1 + \left(\frac{dz(x)}{dx}\right)^2}\,dx. \tag{1.20}$$

Varying (1.20) in respect of the small displacement of the domain boundary $U(x)$, we determine the pressure $p(x)$ acting on the individual sections of the boundary

$$p(x)=\frac{d}{dx}\left(-\frac{\delta\Phi}{\delta U}\right)=$$

$$=\frac{4P_{0z}^2}{\sqrt{\varepsilon_x\varepsilon_z}}\cdot\int\frac{\left[\mathrm{ctg}\psi+\frac{U(x)-U(x')}{x-x'}\right]dx'}{(x-x')\left[\sqrt{\frac{\varepsilon_z}{\varepsilon_x}}+\left(\mathrm{ctg}\psi+\frac{U(x)-U(x')}{x-x'}\right)^2\right]}+$$

$$+\gamma\left[\frac{1}{\sqrt{1+(\mathrm{ctg}\psi+U')^2}}-\frac{(\mathrm{ctg}\psi+U')^2}{\left(\sqrt{1+(\mathrm{ctg}\psi+U')^2}\right)^3}\right]\cdot U''. \tag{1.21}$$

Assuming that the boundary rotates through the small angles $U'(x)\ll 1$ and expanding the integrand in (1.21) in respect of this parameter, we obtain

$$p(x)=\frac{4P_{0z}^2}{\sqrt{\varepsilon_x\varepsilon_z}}\cdot\frac{\left(\frac{\varepsilon_z}{\varepsilon_x}-\mathrm{ctg}^2\psi\right)}{\left(\frac{\varepsilon_z}{\varepsilon_x}-\mathrm{ctg}^2\psi\right)^2}\int\frac{U(x)-U(x')}{(x-x')^2}dx'+$$

$$+\gamma\left|\sin\psi\right|^3\frac{d^2U(x)}{dx^2}. \tag{1.22}$$

The flat domain wall losses the stability with regard to small displacements under the condition of vanishing of the pressure, acting on the wall. Taking into account the low value of the derivative $d^2U/dx^2$ at the moment of the loss of stability, it can be seen that it takes place under the condition of vanishing of the first term in (1.22), i.e. at

$$\mathrm{ctg}^2\psi_c=\frac{\varepsilon_z}{\varepsilon_x}. \tag{1.23}$$

Taking into account that, according to (1.17) at low $\psi$ $\varepsilon_z\approx\frac{4\pi}{2\alpha_z-3\alpha_x\sin^2\psi}$, from (1.23) we obtain the expression for the critical angle of the boundary inclination, at which it losses the stability of its shape [123–131]:

$$\psi_c^2 = \alpha_z / 2\alpha_x. \tag{1.24}$$

For specific experimental conditions there can be a situation, in which the inclination angle of the boundary $\psi > \psi_c$. In this case, the period of the resultant zig-zag structure depends on the extent by which $\psi$ is greater than $\psi_c$. To determine it, let us go over to Fourier components of the wall displacement and of the pressure acting on the wall $U = U_0 \cdot e^{ikx}$, $p = p_0 \cdot e^{ikx}$. Substitution of these into (1.22) gives

$$p_0 = \frac{4P_{0z}^2}{\sqrt{\varepsilon_x \varepsilon_z}} \cdot \frac{\left(\dfrac{\varepsilon_z}{\varepsilon_x} - c\,\mathrm{tg}^2\psi\right)}{\left(\dfrac{\varepsilon_z}{\varepsilon_x} + c\,\mathrm{tg}^2\psi\right)^2} \cdot U_0 \cdot \pi |k| - \gamma |\sin\psi|^3 k^2 \xi_0. \tag{1.25}$$

The competition between the volume and surface energies leads to the period of the boundary $\lambda^*$ for which the rate of the breaking of its flat shape is maximum. It is found from the condition $dp_0/dk = 0$ and is represented by the following equation

$$\lambda^* = \frac{2\pi}{k^*} = \frac{\gamma\sqrt{\varepsilon_x \varepsilon_z}\,|\sin\psi|^3}{P_{0z}^2 \left(\dfrac{\varepsilon_z}{\varepsilon_x} - c\,\mathrm{tg}^2\psi\right)} \left(\frac{\varepsilon_z}{\varepsilon_x} + c\,\mathrm{tg}^2\psi\right)^2. \tag{1.26}$$

For angles close to $\psi$ we obtain

$$\lambda^* = \frac{2\pi\gamma}{P_{0z}^2}\sqrt{\frac{\alpha_z}{2\alpha_x^2}} \cdot \frac{1}{(\psi - \psi_c)}. \tag{1.27}$$

Thus, with the increase of the angle between the plane of the domain wall and the plane, corresponding to its critical inclination, the period of the resultant zig-zag structure decreases.

### 5.2 Broadening of the domain wall as a result of thermal fluctuations of its profile

The deformation of the profile of the domain wall can be not of equilibrium nature but of entropy one caused by thermal fluctuations of its profile. To determine the spectrum of fluctuations of the profile of the domain boundary in the ferroelectric–ferroelastic crystal, let us write down the supplement to the thermodynamic potential of the crystal containing an isolated domain wall with the deformed profile in the form of the functional of the displacement of the boundary $U(z,y)$:

$$\Phi = \int\left\{\frac{\gamma}{2}(\nabla U)^2 + 2P_0\left.\frac{\partial\varphi}{\partial z}\right|_{x=0} U - 2\epsilon_0\sigma_{12}\big|_{x=0} U\right\}d\rho. \tag{2.1}$$

The crystal will be assumed to be infinite and the polar direction and the direction of the spontaneous shear coincide as usual with the $z$ and $y$ directions respectively.

The first term in (2.1) describes the increase of the energy linked with the increase of the area of the domain wall, the second and third describe respectively the energy of the depolarizing field of the bound charges and the elastic energy of the twinning dislocations, formed at the domain wall bending.

The variation (2.1) $\delta\Phi/\delta U$ results in the equation of equilibrium of the boundary

$$-\gamma\nabla^2 U + 2P_0\left.\frac{\partial\varphi}{\partial z}\right|_{x=0} - 2\epsilon_0\sigma_{12}\big|_{x=0} = 0. \tag{2.2}$$

Expanding displacement $U$, potential $\psi$ and the component of the stresses tensor $\sigma_{12}$ into Fourier series:

$$U = \sum_{\mathbf{k}} U_{\mathbf{k}}\cdot e^{ik\rho},\quad \varphi = \sum_{\mathbf{k}} \varphi_{\mathbf{k}}(x)\cdot e^{ik\rho},$$
$$\sigma_{12} = \sum \tilde{\sigma}_{12}(x)\cdot e^{ik\rho},\quad \mathbf{k} = (k_z, k_y),\quad \rho = (z, y), \tag{2.3}$$

we rewrite the expression for $\Phi$ (2.1) in the form of:

$$\Phi = S\sum_{\mathbf{k}}\left\{\frac{\gamma}{2}U_{\mathbf{k}}U_{-\mathbf{k}} + 2P_0 ik_z\varphi_{\mathbf{k}}\big|_{x=0}\cdot U_{-\mathbf{k}} - \right.$$
$$\left. -2\epsilon_0\tilde{\sigma}_{12}\big|_{x=0}\cdot U_{-\mathbf{k}}\right\}. \tag{2.4}$$

As shown in the previous chapter (section 4.1 and 4.3, respectively), the contribution to the Fourier component $\varphi_{\mathbf{k}}|_{x=0}$ and $\tilde{\sigma}_{12}|_{x=0}$, associated with the bending displacement of the boundary, in the approximation of small displacements of the boundary is expressed in the linear form by the Fourier component of its displacement $U_{\mathbf{k}}$:

$$\varphi_{\mathbf{k}}\big|_{x=0} = -\frac{4\pi P_0 ik_z U_{\mathbf{k}}}{\varepsilon_a\left(\frac{\varepsilon_c}{\varepsilon_a}k_z^2 + k_y^2\right)^{1/2}} \tag{2.5}$$

and similarly

$$\tilde{\sigma}_{12}\Big|_{x=0} = -\frac{\mu\epsilon_0}{k}\cdot U_{\mathbf{k}}\left(k_z^2+\omega k_y^2\right), \quad \omega = \frac{2(\lambda+\mu)}{\lambda+2\mu}. \tag{2.6}$$

Substituting into (2.4) $\varphi_{\mathbf{k}}|_{x=0}$ in the form of (2.5) and $\tilde{\sigma}_{12}|_{x=0}$ in the form of (2.6) taking into account the condition $U_{-\mathbf{k}} = U^*_{\mathbf{k}}$, that follows from the reality of displacements $U$, we have

$$\Phi = S\sum_{\mathbf{k}}\left\{\frac{\gamma}{2}k^2 + \frac{8\pi P_0^2 k_z^2}{\varepsilon_a\left(\frac{\varepsilon_c}{\varepsilon_a}k_z^2+k_y^2\right)^{1/2}} - \frac{2\mu\epsilon_0^2\left(k_z^2+\omega k_y^2\right)}{k}\right\}\times$$
$$\times\left|U_{\mathbf{k}}\right|^2 \equiv \frac{S}{2}\sum_{\mathbf{k}}\left|U_{\mathbf{k}}\right|^2\cdot\varphi_{\mathbf{k}}. \tag{2.7}$$

At constant temperature and the volume of the body, the minimum work required for upsetting the equilibrium is equal to the variation of its thermodynamic potential $\Phi$. Therefore, the probability of fluctuations of the domain wall profile is

$$\omega = A'\exp(-\Phi/T). \tag{2.8}$$

Since each of the terms of sum (2.7) depends only on a single $U_{\mathbf{k}}$, the fluctuations of different $U_{\mathbf{k}}$ are statistically independent and their distribution is given by the expression

$$\omega_{\mathbf{k}} = A\cdot\exp\left\{-\frac{S}{2T}\varphi(\mathbf{k})\left|U_{\mathbf{k}}\right|^2\right\} \equiv A\cdot\exp\left\{-\frac{\beta}{2}\left|U_{\mathbf{k}}\right|^2\right\}. \tag{2.9}$$

From the condition of normalization

$$A = \left(\frac{\beta}{2\pi}\right)^{1/2}, \tag{2.10}$$

and then the value of the mean-square fluctuation is

$$\overline{\left|U_{\mathbf{k}}\right|^2} = \sum_k \left|U_{\mathbf{k}}\right|^2\cdot A\cdot\exp\left(-\frac{\beta}{2}\left|U_{\mathbf{k}}\right|^2\right) = \frac{1}{\beta} \equiv \frac{T}{S}\varphi^{-1}(\mathbf{k}). \tag{2.11}$$

On the basis of (2.11) the mean square of the domain wall displacement

$$\overline{U^2} = \sum_{\mathbf{k}}\overline{\left|U_{\mathbf{k}}\right|^2}, \tag{2.12}$$

which can be regarded as a square of a new 'effective' width of the domain wall within which the order parameter changes from its value in one domain to the value in another domain at temperatures different from zero, turns out to be as follows [132]: for the ferroelectric–ferroelastic

$$\overline{U^2} = \frac{T\sqrt{\tilde{\varepsilon}}}{8\sqrt{\pi\mu\epsilon_0}P_0\delta}, \quad \tilde{\varepsilon} \equiv \varepsilon_c = \varepsilon_a, \tag{2.13}$$

for the 'pure' ferroelectric

$$\overline{U^2} = \frac{T\sqrt{\tilde{\varepsilon}}}{2\sqrt{2}\pi P_0\sqrt{\gamma\delta}} \tag{2.14}$$

and, finally, for the 'pure' ferroelastic

$$\overline{U^2} = \frac{T}{4\mu\epsilon_0^2\delta}. \tag{2.15}$$

Numerical evaluations, obtained on the basis of expressions (2.13)–(2.15), show that a considerable ($\overline{U^2} > \delta^2$, $\delta$ is the width of the domain wall with fluctuations not taken into account) broadening of the domain boundaries as the result of thermofluctuations of their profile can appears already at temperatures higher than 100 K.

## 5.3. Effective width of the domain wall in real ferroelectrics

As can be seen from the discussions in chapters 1 and 2, the width of the domain wall in the ferroelectric crystals under theoretical consideration using both phenomenological and microscopic approaches is extremely narrow and close to the lattice constant. At the same time, in the experiments the width of the transition layer between the domains is usually observed being equal to tens or even hundreds of lattice constants [133–140].

The domain structure of a real crystal is to a large extent determined by the nature and type of distribution of its defects. In such a crystal, the interaction of the domain boundary with defects of the crystalline lattice results in deformation of its shape and, consequently, as in the case of thermofluctuations of the profile of the domain wall, the effective thickness of the transition layer between the domains increases. It is natural to assume that one of the possible reasons for the formation of relatively wide domain boundaries observed in the experiments is the fact that in the majority of the experiments recordings were made not of the local thickness of the domain wall which remains narrow, but of the effective transition layer between the domains which forms as a result of deformation of the profile of the domain wall in crystals with defects. This assumption is supported by the experiments based on examination of the width of the boundary in gadolinium molybdate with the help of electron microscope [137–139], which having recorded the value of the thickness of the boundary equal to only several constants of the lattice, differ from the data obtained by optical measurements of the width of the boundary by two or three orders of magnitude.

Let us consider the above assumptions in greater detail. The bending of the domain boundary (deviations of the boundary from the plane of its equilibrium orientation in a defect-free crystal) can be caused not only by the influece of the external field on the domain boundary, pinned by the defects, but also by attraction of the boundary by stationary defects located in the vicinity of the plane describing the position of the middle of the boundary in the crystal. Pinning of a bent boundary by such relatively strong defects, located in accordance with the random nature of their distribution on opposite sides of the plane of average orientation, results in deformation of the shape of the boundary, under which any polar section of the boundary no longer represents a straight line but, in the simplest case, is a curve similar to a broken line.

When considering the geometrically regular, undeformed domain wall in a ferroelectric, the width of the wall is understood as a region within which the polarization vector reverses between its values in the adjacent domains. In crystals with defects, the metioned polarization vector alteration takes place on the average in the layer where the domain boundary is located, bent locally due to its interaction with the defects. The width of this layer is naturally referred to as the effective width of the domain boundary $l_{ef}$ in a real ferroelectric (Fig.5.5). Below, the value of $l_{ef}$ is determined for different concentrations of the point and linear defects, pinning the boundary, at different orient-

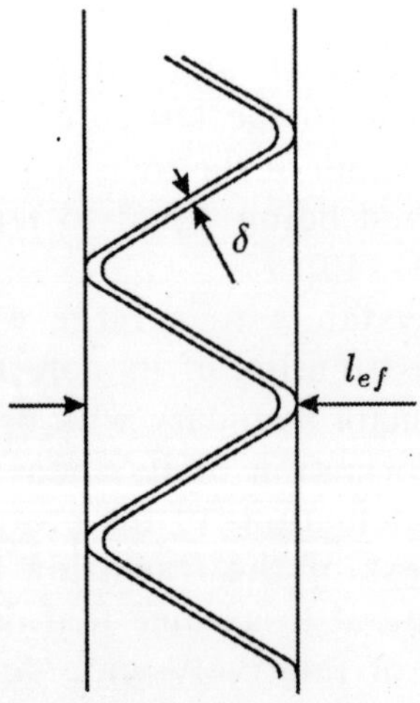


**Fig.5.5.** Increase of the effective length $l_{ef}$ of the transition layer between domains in comparison with the local thickness of the domain wall in a real material.

ation of the latter defects with respect to the polar direction.

As we saw in chapter 4, the interaction of the domain boundaries with crystalline lattice defects creates a potential well for the domain wall in the location of the defect. The deviation of the wall from the bottom of the well by the distance $x$ creates the force $W(x)$ acting on the wall and equal to the derivative along the given direction from the given potential relief taken with the reversed sign. For point defects, the influence of the given force on the boundary is localized not only in the direction normal to the boundary but also in the plane of the domain wall. Consequently the pressure on the boundary from the direction of an individual defect, pinning the boundary, can be presented in the form

$$F = W(x)\cdot\delta(z,y). \tag{3.1}$$

Let us select the origin of the coordinate in such a manner that it coincides not with the defect but with the position of the undisplaced boundary. The profile of the boundary in the vicinity of the defect, pinning the boundary, will be determined by the set of equations

$$\begin{cases} \varepsilon_c \dfrac{\partial^2\varphi}{\partial z^2} + \varepsilon_a\left(\dfrac{\partial^2\varphi}{\partial y^2} + \dfrac{\partial^2\varphi}{\partial x^2}\right) = 8\pi P_0\delta(x)\dfrac{\partial U}{\partial z}, \\ -\gamma\left(\dfrac{\partial^2 U}{\partial z^2} + \dfrac{\partial^2 U}{\partial y^2}\right) + 2P_0\dfrac{\partial\varphi}{\partial z}\Big|_{x=0} = W\cdot\delta(z,y). \end{cases} \tag{3.2}$$

The solution of set (3.2) by the method of the two-dimensional Fourier expansion of the displacement of the boundary and electrostatic potential $\varphi$ in plane $zy$ gives the following equations for the coefficients of expansion $U_{\mathbf{k}}$ and $\varphi_{\mathbf{k}}$:

$$U_{\mathbf{k}} = \frac{1 - 2P_0 ik_z\varphi_{\mathbf{k}}(x=0)}{\gamma(k_z^2 + k_y^2)}, \tag{3.3}$$

$$\varphi_{\mathbf{k}}\big|_{x=0} = -\frac{4\pi P_0 ik_z U_{\mathbf{k}}}{\varepsilon_a\left(\dfrac{\varepsilon_c}{\varepsilon_a}k_z^2 + k_y^2\right)^{1/2}}\cdot\exp\left\{-|x|\left(\frac{\varepsilon_c}{\varepsilon_a}k_z^2 + k_y^2\right)^{1/2}\right\}. \tag{3.4}$$

Expressing $\varphi_{\mathbf{k}}(x=0)$ from (3.3) in terms $U_{\mathbf{k}}$ and equating it to the pre-exponential multiplier in (3.4), we obtain the equation for determination of $U_{\mathbf{k}}$ from which we find [141]:

$$U_{\mathbf{k}} = \frac{W\left(\frac{\varepsilon_c}{\varepsilon_a}k_z^2 + k_y^2\right)^{1/2}}{\left[\gamma k^2\left(\frac{\varepsilon_c}{\varepsilon_a}k_z^2 + k_y^2\right)^{1/2} + 8\pi P_0^2 k_z^2/\varepsilon_a\right]}, \quad k^2 = k_z^2 + k_y^2. \tag{3.5}$$

Using the inverse Fourier transformation from (3.5) for the most typical situation $\varepsilon_c \gg \varepsilon_a$ we obtain

$$U(y,z) = \frac{W}{\gamma}\sqrt{\frac{\lambda}{2\pi z}} \times$$
$$\times\left\{\left[\frac{1}{2} - S(p)\right]\cos\left(\frac{\pi}{2}p^2\right) - \left[\frac{1}{2} - C(p)\right]\sin\left(\frac{\pi}{2}p^2\right)\right\},$$
$$p = \frac{y}{\sqrt{2\pi\lambda z}}, \quad \lambda = \frac{\gamma\sqrt{\varepsilon_c\varepsilon_a}}{8\pi P_0^2}. \tag{3.6}$$

Here $C(p)$, $S(p)$ are Frenel integrals. When deriving (3.6) it was considered that for all $k_z \le 1/\delta$ that are active in bending (it should be remembered that in accordance with the notations used $\delta$ is the thickness of the domain boundary), taking into account the link $\lambda = \gamma\sqrt{\varepsilon_c\varepsilon_a}/8\pi P_0^2 = \sqrt{\varepsilon_a/\varepsilon_c}\,\delta$, the condition $k_z\lambda \ll 1$ is satisfied.

For further considerations, the ratio in the braces in (3.6) can be conveniently presented in the form of a polynomial [98], then the coordinate dependence of the boundary displacement (3.6) is written in the form

$$U(y,z) = \frac{W\sqrt{\varepsilon_c\varepsilon_a}}{8\pi P_0^2} \cdot \frac{\left(\sqrt{2\pi\lambda z} + ay\right)}{\left(4\pi\lambda z + by\sqrt{2\pi\lambda z} + cy^2\right)}, \tag{3.7}$$

where $a$, $b$, $c$ are the numerical coefficients close to respectively 1, 2 and 3. From equation (3.7) it is clearly seen that the localization of the bound charge $\sigma(z,y) = 2P_0\partial U/\partial z$ on the bent boundary along the polar direction $z$ in the vicinity of the pinning point makes the displacement of the boundary sharply anisotropic with the characteristic laws of the decreasing $U \sim 1/\sqrt{z}$ and $U \sim 1/y$ along the polar and non-polar axes, respectively. The analysis of the displacement of the boundary on the basis of (3.5) in the case of $\varepsilon_a > \varepsilon_c$ also confirms the anisotropy of displacement of the boundary, although the law of its decline in this case is different: $U \sim 1/z$ and

$U \sim 1/y^2$ , respectively. At $\varepsilon_a = \varepsilon_c \equiv \tilde{\varepsilon}$ the laws of decrease of the pinned boundary displacement along the polar and non-polar axes are described by the functions $U \sim 1/z^{1/3}$ and $U \sim 1/y^{1/2}$, respectively.

If either of the dielectric permittivity $\varepsilon_c$ or $\varepsilon_a$ is especially high, the displacement of the boundary is controlled completely by surface tension and turns out to be isotropic:

$$U = \frac{W}{2\pi\gamma} \ln\left(l/\sqrt{\pi}\rho\right). \tag{3.8}$$

Hereinafter $l$ is the mean distance between the defects pinning the boundary.

Let us begin the determination of the value of $l_{ef}$ in the case of a ferroelectric with $\varepsilon_a = \varepsilon_c \equiv \tilde{\varepsilon}$ . Neglecting the small displacement of the boundary of the order of the radius of its interaction with defect $a$ (Fig.5.6) in relation to the bottom of the potential well, created by it for the boundary, for the maximum displacement of the boundary on the basis of (3.5) we obtain

$$U_{\max} = \frac{W\sqrt{\tilde{\varepsilon}}}{4\sqrt{2}\pi P_0 \sqrt{\gamma a}}. \tag{3.9}$$

Let us determine the value $W$ at which the domain boundary detaches from the defects. For this purpose, it is necessary to equate the increase of the surface and electrostatic energies, connected to the bending of the boundary, to the energy of interaction of the domain boundary with a defect $\mathcal{U}_0$. The aforementioned increase of the energy is equal to the work by a force $W$ alone, when the wall is displaced. Thus, the condition $\frac{1}{2}\tilde{W} \cdot U_{\max} = \mathcal{U}_0$ taking (3.9) into account, implies that the force of detachment of the wall is

$$\tilde{W} = \left( \frac{8\sqrt{2}\pi P_0 \sqrt{\gamma a}\,\mathcal{U}_0}{\sqrt{\tilde{\varepsilon}}} \right)^{1/2}. \tag{3.10}$$

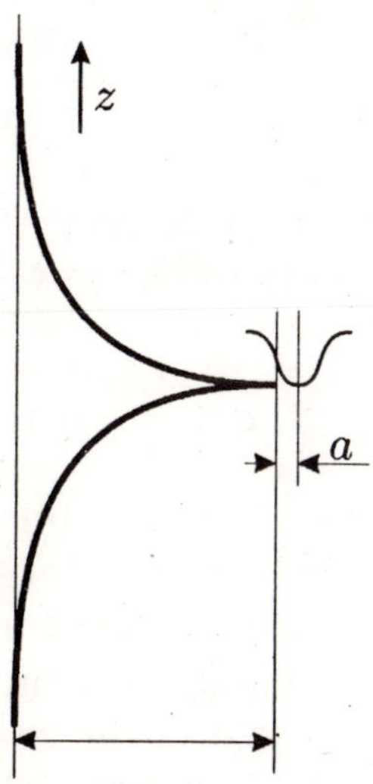


**Fig.5.6.** Profile of the domain wall in the vicinity of the capture of the wall by a point defect.

The average displacement of the

boundary U according to (3.5) is

$$\bar{U} = \frac{W\tilde{\varepsilon}}{8\sqrt{2}\pi^2 P_0^2 l}. \tag{3.11}$$

Substituting in (3.11) $\tilde{W}$ in the form of (3.10) for the effective width of the boundary $l_{ef} = 2\bar{U}\left(\tilde{W}_{OTP}\right)$, in which the average distance between the pinning points $l$ is expressed by the volume concentration of the defects $n$ from the self-consistency condition $\tilde{U} \cdot l^2 = n^{-1}$, where

$$\tilde{U} = U_{\max}\left(\tilde{W}\right) = \left(\frac{\mathcal{U}_0\sqrt{\tilde{\varepsilon}}}{2\sqrt{2}\pi P_0\sqrt{\gamma a}}\right)^{1/2}, \tag{3.12}$$

we obtain the following

$$l_{ef} = \left(\frac{\sqrt{\gamma a}\cdot\tilde{\varepsilon}^{7/2}\cdot n^2\mathcal{U}_0^3}{2^3\left(\sqrt{2}\right)^3\pi^7 P_0^7}\right)^{1/4} \tag{3.13}$$

At conventional $\mathcal{U}_0 \sim (T-T_c)^{3/2}$ the value of $l_{ef}$ is proportional to $(T-T_c)^{1-2}$, decreasing, in contrast to $\delta$, when the phase transition point is approached.

If the displacement of the boundary is determined by the equation (3.8), the effective width of the domain wall, obtained from identical considerations, is determined by the expression

$$l_{ef} = \sqrt{\frac{\mathcal{U}_0}{2\pi\gamma}\ln^{-1}\left(\sqrt{\pi\gamma}/2na^2\sqrt{\mathcal{U}_0}\right)}. \tag{3.14}$$

When domain boundaries are pinned by linear defects whose axes are perpendicular to the vector $\mathbf{P}_0$, similarly to (3.5) we have

$$U_k = \frac{W_\tau\sqrt{\varepsilon_c\varepsilon_a}}{\left(\gamma k^2\sqrt{\varepsilon_c\varepsilon_a} + 8\pi P_0^2 k\right)}, \tag{3.15}$$

where $W_\tau$ is the average force acting on the boundary from the direction of the unit of length of the linear defect.

As mentioned above, the value $\gamma\sqrt{\varepsilon_c\varepsilon_a}/8\pi P_0^2$ is always lower than $1/k$ and, consequently, the first term in the denominator of (3.15) can be ignored for all real $k$. It means that in the case of the pinning of the domain boundary by linear defects of the mentioned orientation its profile is completely determined by the interaction of the bound charges that occur at bending of the boundary on its surface, and turns out to be as follows

$$U(z)=\frac{W_\tau\sqrt{\varepsilon_c\varepsilon_a}}{8\pi P_0^2}\cdot\ln\frac{l}{2z},\tag{3.16}$$

From the conditions $\frac{1}{2}\cdot U_{max}\tilde{W}_\tau=\mathcal{U}_{0\tau}$, where $\mathcal{U}_{0\tau}$ is the energy of interaction with the unit of length of the defect, taking into account (3.16), the linear density of the detachment force is

$$\tilde{W}_\tau^{OTP}=\frac{4\pi P_0\sqrt{\mathcal{U}_{0\tau}}}{(\varepsilon_c\varepsilon_a)^{1/4}\sqrt{\ln l/2a}}.\tag{3.17}$$

On the basis of the ratio $l\cdot\tilde{U}=n_s^{-1}$, where $n_s$ is the surface density of linear defects and $\tilde{U}=U_{max}\left(\tilde{W}_\tau\right)$,

$$l_{ef}\simeq\frac{P_0}{n_s(\varepsilon_c\varepsilon_a)^{1/4}\sqrt{\mathcal{U}_{0\tau}}}.\tag{3.18}$$

and consequently

$$l_{ef}=2U\left(\tilde{W}_\tau\right)=\frac{\tilde{W}_\tau\sqrt{\varepsilon_c\varepsilon_a}}{4\pi^2P_0^2}=\frac{(\varepsilon_c\varepsilon_a)^{1/4}\sqrt{\mathcal{U}_{0\tau}}}{\sqrt{\ln\left(P_0/n_sa(\varepsilon_c\varepsilon_a)^{1/4}\sqrt{\mathcal{U}_{0\tau}}\right.}}.\tag{3.19}$$

For pinning the domain boundaries by linear defects, whose axes are parallel to the polar direction

$$U_{max}=W_\tau\cdot l/4\gamma.\tag{3.20}$$

In this case the density of the detachment force is

$$\tilde{W}_\tau=\sqrt{8\gamma\mathcal{U}_{0\tau}/l}.\tag{3.21}$$

The average distance between the defects, pinning the boundary, is

$$l=\left(\frac{\sqrt{2\gamma}}{n_s\sqrt{\mathcal{U}_{0\tau}}}\right)^{2/3}.\tag{3.22}$$

and, finally

$$l_{ef}=2U\left(\tilde{W}_\tau\right)=\tilde{U}=\left(\frac{\mathcal{U}_{0\tau}}{2n_s\gamma}\right)^{1/3}.\tag{3.23}$$

It should be noted that equations (3.19), (3.23) can be used for the case of oriented axes of linear defects, formed under specific conditions of crystal preparation. In the case of the arbitrary orientation of the axes of these defects they are usually intersected with the plane of the domain boundary. In this case, the pinning of the boundary by defects is more similar to the case of point pinning and it appears that equations (3.13), (3.14) are more suitable for determining $l_{ef}$.

Numerical estimates of the value $l_{ef}$ at $P_0 \sim 10^4$, $\mathcal{U}_0 \sim 1$ eV, $a \sim 10^{-7}$ cm, $\mathcal{U}_{0\tau} \sim \mathcal{U}_0/a \sim 10^{-5}$ ($a$ is the size of the elementary cell), $\gamma \sim 1$ erg/cm$^2$ give the following results. For a point defect with $n \sim 10^{18}$ cm$^{-3}$, $l_{ef} \simeq 4 \cdot 10^{-7}$ cm in the absence of compensation of long-range forces and $l_{ef} \sim 10^{-6}$ cm in the presence of such a compensation. For linear defects with $n_s \sim 10^8$ cm$^{-2}$ in the case when their axes are perpendicilar to the vector of spontaneous polarization, $l_{ef}^{\perp} \simeq 10^{-6}$ cm, otherwise at the same value $n_s$, $l_{ef}^{\parallel} \leq 10^{-4}$ cm. These estimates show that for all types of defects at their real concentration, the value of $l_{ef}$ is greater or considerably greater than $\delta$. This allows us to assume that the observation in experiments of wide domain walls with the thickness considerably greater than $\delta$ can be attributed to the interaction of domain boundaries with crystal defects.

This is also proved by the fact that temperature dependence of $l_{ef}$ differs in comparison with the prediction of the standard thermodynamic theory (equation (1.21) in chapter 2). The temperature dependence of $l_{ef}$ is closer, for example, to the experimental results, obtained by measurements of the thickness of the domain wall in triglycine sulphate crystal [134] where the decrease of the domain wall thickness at $T \to T_c$ (Fig 5.7) is observed instead of its increase.

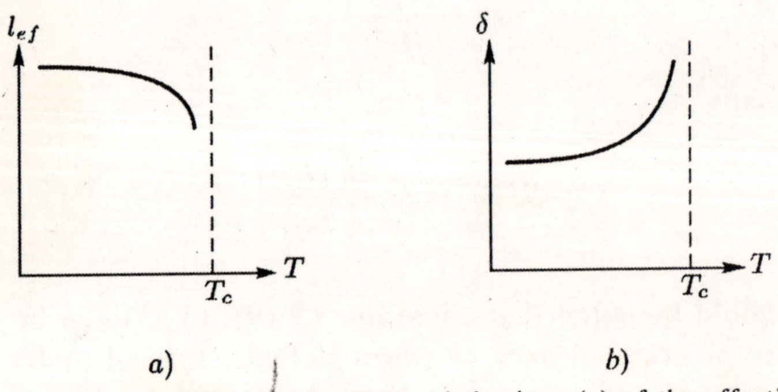


**Fig. 5.7.** Qualitatively different temperature behaviour (a) of the effective and (b) local thickness of the domain wall.

## 5.4 Effective width of the domain wall in ferroelastic with defects

The discussion of the width of the transition layer between the domains in a ferroelectric, caried out in the previous section, did not take into account the change of the elastic energy of the crystal at deformation of the shape of the domain wall. Such a consideration describes 'pure' ferroelectrics, for example a TGS crystal which, according to Aizu classification, is not a ferroelastic. At the same time, a large number of ferroelectrics also undergo ferroelastic deformation during phase transitions. In addition there are the so-called 'pure' ferroelastics, which completely lack ferroelectric properties. When considering the structure of the deformed domain wall in all such crystals, it is necessary to take into account elastic effects.

Let us determine the form of the domain boundary interacting with the defect and the effective width of the domain wall in a ferroelastic. The displacement of the boundary interacting with the defects is determined similarly to 5.3 from the compatible solution of the set of equations, one of which, as before, is the equation of equilibrium of the boundary and the role of the other one is played by the condition of incompatibility of elastic strains written for the static case:

$$\begin{cases} -\gamma\left(\dfrac{\partial^2 U}{\partial z^2}+\dfrac{\partial^2 U}{\partial y^2}\right)-2\epsilon_0\sigma_{12}\big|_{x=0}=W\delta(z,y), \\ \sigma_{ij,kk}+\dfrac{m}{m+1}\left(\sigma_{kk,ij}-\sigma_{kk,ll}\delta_{ij}\right)=2\mu\eta_{ij}. \end{cases} \tag{4.1}$$

As previously, the considered material is assumed to be isotropic in respect of elasticity.

The connection of the components of the tensor of elastic stresses with the displacement of the domain wall determined by the equation of incompatibility of the strain in (4.1) by the dependence of tensor $\eta_{ij}=\eta_{ij}(U)$ on the wall displacement naturally turns out to be the same as in the previous problems (section 4.3, 5.2) that dealt with the bending displacement of the walls in elastics. In particular, the Fourier image

$$\begin{gathered} \tilde{\sigma}_{12}(x=0)=-\frac{\mu\epsilon_0}{k}U_{\mathbf{k}}\left(k_z^2+\omega k_y^2\right), \\ \omega=2(\lambda+\mu)/(\lambda+2\mu). \end{gathered} \tag{4.2}$$

The equation of the boundary equilibrium (4.1) in the Fourier space has the form

$$\gamma\left(k_z^2+k_y^2\right)\cdot U_{\vec{k}}-2\epsilon_0\tilde{\sigma}_{12}\big|_{x=0}=W. \tag{4.3}$$

Hence, taking into account (4.2)

$$U_{\mathbf{k}}=\frac{W\cdot k}{\left[\gamma k^3+2\mu\epsilon_0^2\left(k_z^2+\omega k_y^2\right)\right]}. \tag{4.4}$$

The analysis of the original obtained on the basis of the Fourier image (4.4) shows that for almost all $\rho$ only low values of $k$ are active in the displacement of the boundary and in this case

$$\gamma k^3 \ll 2\mu\epsilon_0^2\left(k_z^2+\omega k_y^2\right) \tag{4.5}$$

and consequently, only the second term can be left here in the denominator (4.4). In this case, the coordinate dependence of the displacement of the boundary has the form [142]:

$$U(z,y)=\frac{W}{4\pi\mu\epsilon_0^2}\cdot\frac{\sqrt{z^2+y^2}}{\left[z^2\omega+y^2\right]}, \tag{4.6}$$

i.e. it possesses the characteristic law of decrease $\sim 1/\rho$. Comparison of displacement (4.6) with displacement of the boundary (4.3), determined only by surface tension, shows that the latter in fact determines displacement of the boundary only at $\rho<\gamma/2\mu\epsilon_0^2\cdot\ln\left(l/\sqrt{\pi}a\right)$, i.e. almost beyond the limits of applicability of consideration of $\rho\geq a$ carried out here.

To determine the effective width of the boundary $l_{ef}$ let us first of all find the value of $W$ at which the boundary detaches itself from defects. On the basis of the previously mentioned condition $\frac{1}{2}\cdot U_{max}\cdot\tilde{W}=\mathcal{U}_0$ ($\mathcal{U}_0$ is the energy of interaction of the boundary with the defect) and of equation (4.6) we have

$$\tilde{W}=2\epsilon_0\sqrt{2\pi\mu\mathcal{U}_0 a}. \tag{4.7}$$

The average displacement of the boundary is

$$\overline{U}=\frac{W}{2\sqrt{\pi}\mu\epsilon_0^2\cdot l}. \tag{4.8}$$

According to the conditions $U_{\max}\left(\tilde{W}\right)\cdot l^2=n^{-1}$, the average distance between the defects pinning the boundary is

$$l = \frac{1}{\sqrt{2n}} \cdot \left( \frac{8\pi\mu\epsilon_0^2 a}{\mathcal{U}_0} \right)^{1/4} \tag{4.9}$$

Then, taking into account (4.7)–(4.9), the effective width of the domain wall $l_{ef} = 2\overline{U}\left(\tilde{W}\right)$ turns out to be the following:

$$l_{ef} = \sqrt{\frac{2}{\pi}} \cdot \left( \frac{\mathcal{U}_0}{\mu\epsilon_0^2} \right)^{3/4} \cdot n^{1/2} \left(8\pi a\right)^{1/4} \cdot \tag{4.10}$$

In conclusion of the consideration of the deformed profile of the domain wall in crystals with defects, it is important to note the following. As it follows from the linearity of the equations used in this case, the magnitude of the maximum displacement of the wall in the region of bending increases linearly with the increase of the force acting on the wall. At the same time, the bending itself being controlled by the long-range electrical or elastic fields both in the case of the ferroelectric and the ferroelastic is extremely localized in the vicinity of pinning of the bent wall (Fig.5.8). Consequently, if the displacement of the domain wall counted from the location of the defect is discussed (which is natural, for example, in the problem of displacement of a pinned domain wall in the external field), then for not so high concentration of the defects the average displacement of the wall coincides almost completely with its maximum value. This means that the quantity $\overline{U}$ is also proportional to $W$. Introducing the proportionality coefficient between $\overline{U} = U_{max}$ and $W$ from the condition $W = \vartheta U_{max}$ on the basis of expressions (4.3) and (4.6) we obtain the effective coefficients of the quasi-elastic force, acting on the boundary displaced with regard to the defect, which is pinning it. For a 'pure' ferroelectric

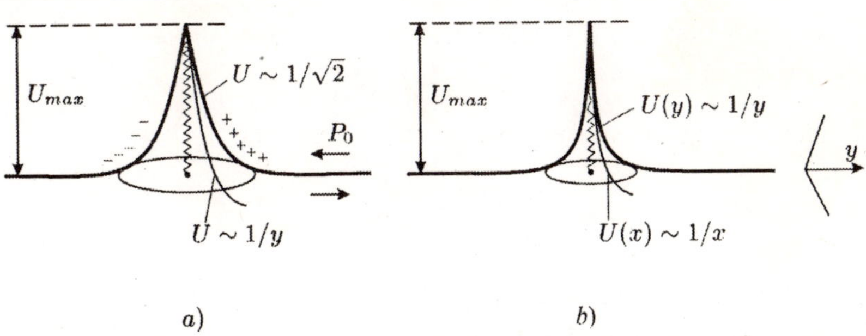


**Fig.5.8.** Localization of the region of bending in the vicinity of pinning the domain wall in the case of (a) ferroelectric and (b) ferroelastic. The closed line shows the lines of the equal displacements of the domain wall.

$$\vartheta = \frac{4\sqrt{2}\pi P_0 \sqrt{\gamma a}}{\sqrt{\tilde{\varepsilon}}}. \tag{4.11}$$

For a 'pure' ferroelastic

$$\vartheta = 4\pi\mu\epsilon_0^2 a. \tag{4.12}$$

At that the domain wall being displaced now is regarded already as a flat one that evidently greatly simplifies further consideration.

In the case of the ferroelectric–ferroelastic, the Fourier image of the boundary displacement is obtained by adding the term $2\pi\mu\epsilon_0^2\left(k_z^2+\omega k_y^2\right)/k$ to the denominator of the expression for $U_{\mathbf{k}}$ (3.5) of the 'pure' ferroelectric. As the result, the structure of displacement of the wall turns out to be qualitatively similar to the case of 'pure' ferroelastic (4.6), i.e. $U \sim 1/\rho$, and the effective coefficient of the quasi-elastic force is:

$$\vartheta = \frac{4\sqrt{\pi\mu}\epsilon_0 P_0 a}{\sqrt{\tilde{\varepsilon}}}. \tag{4.13}$$

## Chapter 6

# Mobility of domain boundaries in crystals with different barrier height in a lattice potential relief

As shown in Chapter 3, the magnitude of the lattice barrier, surmounted by the wall during its motion, strongly depends on the structure of the domain wall, and, in particular, on its width. It will be shown below that a similar dependence of the domain boundary mobility also exists in the cases when the influence of the mentioned relief on the domain wall motion can be ignored.

To study the mobility of domain boundaries in ferroelectrics, we first of all consider the parameters of moving domain boundaries within the framework of the continual approximation.

### 6.1. Structure of the moving boundary, its limiting velocity and effective mass of a domain wall within the framework of the continual approximation. Mobility of the domain boundaries

To determine the parameters of the moving domain wall, the expression (1.10) in Chapter 1 must be supplemented by the density of kinetic energy $T$. Writing explicitly only the ferroactive displacements of the particles, we have

$$T=\frac{1}{2}\rho\left(\frac{\partial u}{\partial t}\right)^2=\frac{1}{2}\mu\left(\frac{\partial P}{\partial t}\right)^2,\quad \mu=\rho a^6/e^{*2}, \qquad (1.1)$$

where $u$ is the displacement of the ferroactive particles leading to the occurrence of polarization $P$, $\rho$ is the density of the crystal, $a$ is the size of the elementary cell, $e^*$ is the effective charge linking $u$ with $P$ and $a$.

Taking into account (1.1), the surface density of the total energy of the ferroelectric is

$$\Phi = \int_{-\infty}^{\infty} \left(\overline{\Phi} + T\right) dx =$$
$$= \int \left\{ \frac{\mu}{2} \left( \frac{\partial P}{\partial t} \right)^2 + \frac{\varkappa}{2} \left( \frac{\partial P}{\partial x} \right)^2 - \frac{\alpha}{2} P^2 + \frac{\beta}{4} P^4 \right\} dx. \tag{1.2}$$

On the basis of (1.2), the equation of motion for polarization in the absence of dissipation and the external effects can be written in the form of

$$\mu \frac{\partial^2 P}{\partial t^2} - \varkappa \frac{\partial^2 P}{\partial x^2} - \alpha P + \beta P^3 = 0. \tag{1.3}$$

Assuming further that the distribution of polarization in the moving wall $P(x,v) = P(x-vt)$, where $v$ is velocity of the domain wall motion, taking into account the consequent ratio between the derivatives $\frac{\partial P}{\partial t} = -v \frac{\partial P}{\partial x}$ in the coordinate system moving together with the wall, where $x' = x - vt$, we can rewrite the equations (1.3) for distribution of polarization in the boundary in the following form

$$\tilde{\varkappa} \frac{\partial^2 P}{\partial x'^2} = -\alpha P + \beta P^3,$$
$$\tilde{\varkappa} = \varkappa - \mu v^2 = \varkappa \left(1 - v^2 / c_0^2\right), \tag{1.4}$$
$$c_0^2 = \frac{\varkappa}{\mu}.$$

Equation (1.4) precisely coincides with the equation (1.6) of chapter 2 with the accuracy up to substitution $\varkappa \to \varkappa$ and $x \to x'$, and, therefore, we immediately write down the distribution of polarization in the moving domain wall as

$$P(x,v) = P_0 \cdot \mathrm{th} \frac{(x - vt)}{\delta \sqrt{1 - v^2 / c_0^2}}, \quad \delta = \sqrt{\frac{2\varkappa}{\alpha}}. \tag{1.5}$$

According to (1.5) there is the limiting velocity of motion of the domain wall $c_0 = \sqrt{\varkappa / \mu}$, approaching which we observe the 'Lorenz' reduction of the width of the moving domain wall $\tilde{\delta} = \delta \sqrt{1 - v^2 / c_0^2}$ as compared to its static value. In the absence of viscosity and of the external field the domain wall can freely move with a permanent velocity, which assumes in magnitude arbitrary values between zero and the limiting value $c_0$.

Let us determine the energy of the moving boundary. Substitution of distribution (1.5) into (1.2), where $\overline{\Phi} \to \overline{\Phi} - \overline{\Phi}(P_0)$ gives

$$\gamma(v) = \frac{4\varkappa P_0^2}{3\delta} \cdot \frac{1}{\sqrt{1 - v^2 / c_0^2}} = \frac{\gamma_0}{\sqrt{1 - v^2 / c_0^2}} = m^*(v) c_0^2. \tag{1.6}$$

Here $\gamma_0$ is the energy of the static domain wall coinciding with expression (1.22) in chapter 2, and

$$m^*(v) = \frac{\gamma_0 / c_0^2}{\sqrt{1\sqrt{1 - v^2 / c_0^2}}} = \frac{m^*}{\sqrt{\sqrt{1 - v^2 / c_0^2}}} \tag{1.7}$$

is the so-called effective mass of the unit area of the domain wall, which at low velocities of the wall $v \ll c_0$ is equal to its limiting value

$$m^* = \frac{\gamma_0}{c_0^2}. \tag{1.8}$$

Substitution in (1.8) of the explicit expression for $\gamma_0$ (1.6), $c_0$ (1.4) and $\mu$ (1.1) taking into account the ratio $P_0 = e^* u_0 / a^3$ enables us to write (1.8) in the form suitable for analysis

$$m^* = \frac{4}{3} \rho \delta \frac{u_0^2}{\delta^2} = \frac{4}{3} \rho \epsilon_0^2 \delta. \tag{1.9}$$

Here $u_0$ is the value of spontaneous displacement of the ferroactive particles, and $\epsilon_0 = u_0 / \delta$ is the deformation, corresponding to this displacement. For the ferroelastic crystals $\epsilon_0$, in particular, is the spontaneous deformation.

Equation (1.9) shows clearly that, with regard to the order of magnitude, the effective mass of the wall is the product of the actual mass of the particles, located within the limits of the domain wall, $\rho\delta$, by the dimensionless multiplier $\epsilon_0^2$. Under typical $\rho \sim 1$ g/cm$^3$, $\delta \sim 10^{-7}$ cm and $\epsilon_0 \sim 10^{-2}$ we have $m^* \sim 10^{-11}$ g·cm$^{-2}$ [143].

When taking into account dissipation and the presence of the external field, the equation of motion for polarization has the form

$$\mu \frac{\partial^2 P}{\partial t^2} + \Gamma \frac{\partial P}{\partial t} - \alpha P + \beta P^3 - \varkappa \frac{\partial^2 P}{\partial x^2} = E. \tag{1.10}$$

To determine the solution of equation (1.10), let us first of all examine its asymptotics. Away from the boundary, where the values of all derivatives are equal to zero, the asymptotic values of polarization are the roots of the equation

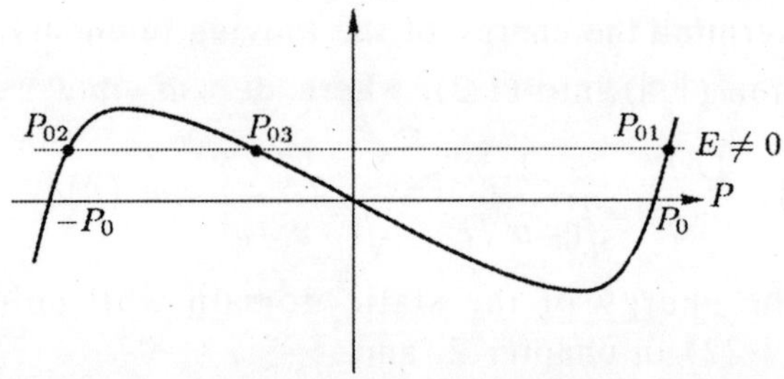


**Fig.6.1.** Roots of the polynomial $-\alpha P+\beta P^3 -E$ at different values of the external field.

$$-\alpha P + \beta P^3 = E. \tag{1.11}$$

At $E \neq 0$, these roots $P_{01}$, $P_{02}$, $P_{03}$ (see Fig 6.1) no longer have those ratios of symmetry $P_{01} = -P_{02}=\alpha/\beta$, $P_{03}=0$ which exist in the crystal in the absence of the external field. In the given case, $P_{01} > P_0$, $|P_{02}| < P_0$ and $P_{03}\neq 0$. In accordance with the definition of the domain boundary, in one limit polarization in the boundary should have the value of $P_{01}$, and in the other limit the value of $P_{02}$, not equal to the former one in magnitude.

To form the solution of equation (1.10) with the mentioned asymptotics, it is convenient to write the equation using the dimensionless variables first $p = P/P_0$, $\xi = \sqrt{2}(x - vt)/\delta\sqrt{1 - v^2/c_0^2}$. Dividing both parts of (1.10) by $\alpha P_0$ we obtain

$$\frac{1}{2}\frac{\delta^2}{c_0^2}\cdot\frac{\partial^2 p}{\partial t^2} + \frac{\Gamma}{\alpha}\frac{\partial p}{\partial t} - p + p^3 - \frac{1}{2}\delta^2\frac{\partial^2 p}{\partial x^2} = E', \tag{1.12}$$

where $E'=E/\alpha P_0$. Therefore, taking into account the relationship

$$\frac{\partial^2 p}{\partial t^2} = \frac{\partial^2 p}{\partial \xi^2}\cdot\frac{2v^2}{\delta^2\left(1 - v^2/c_0^2\right)},$$
$$\frac{\partial^2 p}{\partial x^2} = \frac{\partial^2 p}{\partial \xi^2}\cdot\frac{2}{\delta^2\left(1 - v^2/c_0^2\right)}, \tag{1.13}$$

the equation for the distribution of polarization in the moving boundary in the presence of dissipation is

$$\frac{\partial^2 p}{\partial \xi^2} + \tilde{v}\frac{\partial p}{\partial \xi} + p - p^3 + E' = 0, \tag{1.14}$$

where

$$\tilde{\upsilon} = -\frac{\sqrt{2}\cdot\Gamma}{\alpha\cdot\delta\sqrt{1-\upsilon^2/c_0^2}}\cdot\upsilon\cdot \tag{1.15}$$

To write the solution of equation (1.14) we use the solution of this equation at $\tilde{\upsilon}$, $E'=0$ (ratio (1.21) in chapter 2) written in the asymmetrical form

$$\frac{p}{p_0} = \text{th}\frac{x}{\delta} = \frac{\exp(2x/\delta)-1}{\exp(2x/\delta)+1} = 1 - \frac{2}{\exp(2x/\delta)+1}. \tag{1.16}$$

Taking into account the changed asymptotics, let us find the solution of equation (1.14) in the form of [144,145]

$$p(\xi) = a + \frac{(b-a)}{\left[1+\exp(b-a)\xi/\sqrt{2}\right]}, \tag{1.17}$$

where $a=p_{01}/p_0$, $b=p_{02}/p_0$, $c=p_{03}/p_0$ are the dimensionless roots of the polynomial

$$p^3 - p - E' = (p-a)(p-b)(p-c). \tag{1.18}$$

Substitution of (1.17) into (1.14) shows that function (1.17) is the solution of equation (1.14) at

$$\tilde{\upsilon} = \frac{(a+b-2c)}{\sqrt{2}} = -\frac{3c}{\sqrt{2}}. \tag{1.19}$$

The last ratio follows from the condition $a+b+c=0$, which is satisfied by the roots of the polynomial (1.18) because of the absence of the quadratic term in it.

If we know the root, taking into account (1.15) and (1.19), we obtain an implicit dependence of the velocity of the domain wall $\upsilon$ on the magnitude of the applied external electric field [146–148]. At low velocities of the wall, this dependence can be written in the explicit form. In this case, as it can be seen in Fig.6.2, $c=p_{03}/p_0 \simeq E/\alpha p_0$,

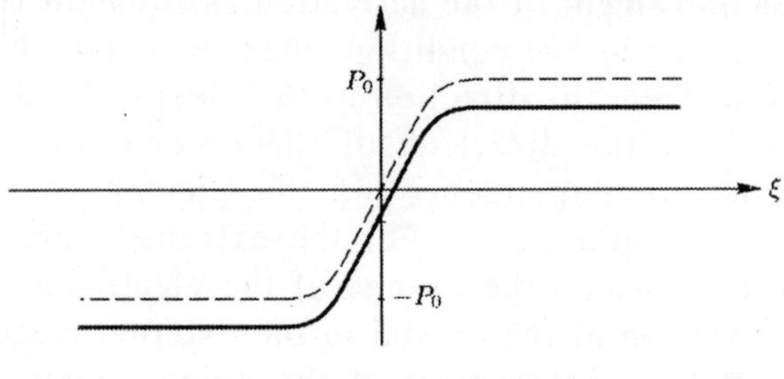


**Fig.6.2.** Distribution of polarization in a stationary domain wall and in a domain wall moving in the external field.

$\upsilon \ll c_0$, and, therefore, according to (1.15) and (1.19) we obtain

$$\upsilon = \frac{3}{2}\frac{E}{P_0}\frac{\delta}{\Gamma} = \mu E, \quad \mu = \frac{3}{2}\frac{\delta}{P_0}\frac{1}{\Gamma}. \tag{1.20}$$

Thus, in weak field the velocity of the wall depends linearly on the field through the mobility $\mu$, determined by the relationship (1.20) [149]. It should be mentioned that with the accuracy up to the numerical coefficient, ratio (1.20) can be written immediately from the initial equation (1.10) assuming that the external field and the dissipation have no influence on the profile of the moving wall. Then assuming that the profile is determined by equation (1.3) and equating subsequently the terms $\Gamma \partial P/\partial t = E$, where $\partial P/\partial t \simeq P_0/\delta$, we immediately obtain ratio (1.20).

It should be also noted that according to (1.17) the distribution of polarization in the wall moving in the external field is determined in any fields only by the asymptotic values of polarization, i.e. depends on the strength of the external field and does not depend on dissipation.

## 6.2. Lateral motion of domain boundaries in ferroelectric crystals with high values of the barrier in the lattice relief of domain walls. The thermofluctuation mechanism of the domain wall motion. Parameters of lateral walls of the critical nucleus on a domain wall

The expressions for the domain wall velocity and its mobility derived above are applicable to relatively wide domain boundaries formed in the vicinity of $T_c$, for which the influence of the lattice relief on their motion can be ignored. For conventional domain walls that are usually narrow the presence of the lattice relief, connected to the coordinate dependence of their energy, almost completely prevents their motion as a unit in relatively weak external fields. In fact, the achievement of the activationless domain wall motion mode is determined by the condition when the external pressure on the domain wall from the direction of the electric field $E_{cr}$ exceeds the pressure from the direction of the Peierls' force $\partial\gamma/\partial U|_{max}$, where $\gamma(U)$ is the dependence of the energy of the domain boundary on its displacement. For the extremely narrow domain wall with zero thickness the change of the electrostatic energy of the dipole subsystem of the crystal in the external electric field $\mathbf{E}$, resulting from the displacement of the domain wall, is equal to $\delta\Phi = 2(\mathbf{P}_0\mathbf{E})\delta U$, where $\delta U$ is the displacement of the wall. Hence, the pressure on the wall from the direction of the external field is

$$p=\frac{\delta\Phi}{\delta U}=2(\mathbf{P}_0,\mathbf{E}). \tag{2.1}$$

Equating the pressure (2.1) to the pressure from the direction of the Peierls' force

$$2(\mathbf{P}_0,\mathbf{E}_{cr})=\left.\frac{\partial\gamma}{\partial U}\right|_{max}\simeq\frac{2V_0}{a}, \tag{2.2}$$

where $V_0$ is the magnitude of the barrier in the lattice relief, $a$ is the size of the elementary cell, we determine the strength of the critical field

$$E_{cr}\simeq\frac{V_0}{P_0 a}. \tag{2.3}$$

Calculations of $V_0$ for certain ferroelectrics, presented in chapter 3 show that, in particular, even for a crystal with a highly mobile domain structure – potassium dihydrophosphate, the values of $V_0$ at $(T_c-T)$ equal to several degrees are equal to the order of several hundredths of erg·cm$^{-2}$. At these values of $V_0$ and $P_0$~$10^4$ of CGSE units, $a$~$10^{-7}$ cm, $E_{cr}$ is of the order of ~1 kV·cm$^{-1}$. In crystals with a less mobile domain structure the value of $E_{cr}$ is expected to be approximately by an order of magnitude greater.

The mentioned estimates correspond to the results of a large number of experiments carried out to determine the inverse switching time of the ferroelectric crystal. As can be seen, in particular, in Fig.6.3, which shows this dependence for the crystal of triglycine sulphate, the curve of switching current can be qualitatively divided into two sections. In section I, the inverse switching time and, consequently, the velocity of the domain boundaries motion follows by the exponential law $1/t_s=1/t_\infty\cdot\exp(-\bar{\delta}/E)$, $\upsilon=\upsilon_\infty\cdot\exp(-\bar{\delta}/E)$. In section II, this dependence follows the linear law: $1/t_s=\text{const}\cdot E$ ($v=\text{const}\cdot E$). The specific value of the critical field $E_{cr}$, separating these sections, for the crystal of triglycine sulphate is ~20 kV·cm$^{-1}$.

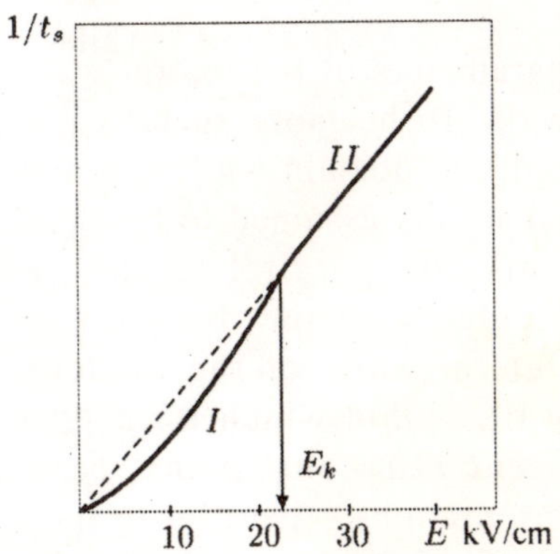


**Fig. 6.3.** Dependence of the inverse switching time on the field for TGS crystal [16].

In fields $E>E_{cr}$, the motion of the

domain boundaries obviously takes place in the activationless way and is described by the dependence $v=\mu E$, obtained in the previous sections of this chapter.

In the fields weaker than $E_{cr}$, the lateral motion of the domain walls as a unit is imaginary. Here it is carried out with high probability by way of formation of nuclei of the inverse domains on the lateral surface of a domain wall with their subsequent growth. A large number of studies [150–156] from Drougard [150], Miller and Weinreich [151] to Hayashi [153,154] were devoted to the development of this concept. A special attention in the most thorough investigations [153,154] was paid to the detailed consideration of the kinetics of the process. However, the initial stage of nucleation is considered in almost all studies [150–154] on the basis of oversimplified modelling consideration (imagining a nucleus having a triangular, squared shape, etc.).

Recently, the equilibrium form of the critical nucleus on the domain walls was determined in [155]. However, even in this case in a number of instances (determination of the energy of the charged section of the lateral wall of the nucleus by way of its replacement with the corresponding section of the dielectric ellipsoid, the application of isotropic approximation to the velocity of motion of the lateral walls of the nucleus of different orientation, etc.) the authors did not use correct enough approximations. In addition to this, as it is shown below, the restriction of the test function type, used in [155] in solving the variation problem, does not give the accurate concept of the shape of the critical nucleus.

The successive determination of the critical nucleus parameters and (on this basis) of velocity of the domain wall in the external field was carried out in [156].

Let's find subsequent to [156] the parameters of the lateral walls of the critical nucleus on a domain wall. To be more specific, let us assume that the plane of the nondisplaced domain wall, and also the flat wall of the nucleus, whose thickness is assumed to be equal to the constant of the elementary cell, are parallel to the $zy$-coordinate plane and the $x$ axis coincides with the direction of displacement of the boundary. The lateral walls of the nucleus, representing sections of the domain wall, within which it changes from some $x$=const plane to the adjacent $x\pm a$=const plane, have, evidently, two qualitatively different orientations – parallel to the $z$ axis which is the direction of the vector of spontaneous polarization, and parallel to the $y$ axis respectively. It will be shown below that the walls of both types have width $\lambda$ much greater than $a$, so that

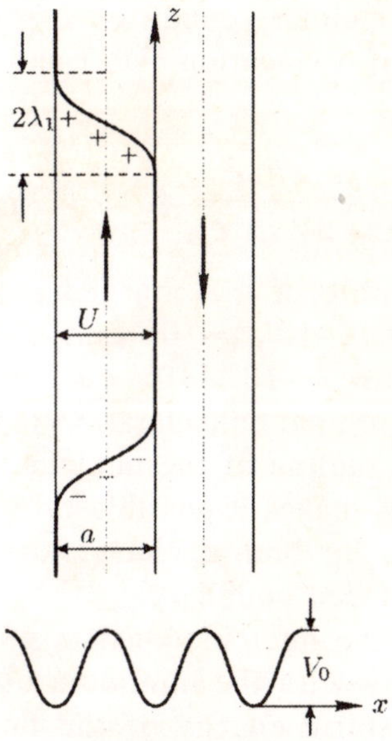


**Fig.6.4.** Formation of side walls of a nucleus on the domain wall in the course of transition from a valley of the lattice relief to the adjacent one. $U$ is the displacement of the wall. $2\lambda_1$ is the width of the charged side wall of the nucleus. Arrows indicate the direction of the vector of polarization in adjacent domains.

the continual approximation can be used when describing them.

The structure of the charged lateral wall of the nucleus, parallel, to the $y$-axis (Fig 6.4), neglecting surface tension, is determined by the condition of equality of the pressure on the boundary from the direction of the field of bound charges on the boundary to the pressure from the direction of the Peierls' force:

$$2P_0\int_{-\infty}^{\infty}\frac{2\sigma(z')dz'}{\sqrt{\varepsilon_c\varepsilon_a}\cdot(z-z')}=-\frac{d\gamma}{dU}. \tag{2.4}$$

Here $\sigma(z')$ it the density of the bound charges on the boundary. The integral in (2.4) has the meaning of the main value in order to exclude the physically meaningless action of the bound charge on itself.

Substituting in (2.4) the density of the bound charge on the boundary, expressed with the help of the boundary displacement

$$\sigma(z')=2P_0 dU(z')/dz', \tag{2.5}$$

for the simplest form of the periodic potential relief

$$\gamma(U)=\frac{V_0}{2}\cos\frac{2\pi U(z)}{a} \tag{2.6}$$

equation (2.4) can be written in the form

$$\frac{8P_0^2}{\sqrt{\varepsilon_c\varepsilon_a}}\int_{-\infty}^{\infty}\frac{dU(z')}{dz'}\cdot\frac{dz'}{(z-z')}=-\frac{V_0\pi}{a}\cdot\sin\left(\frac{2\pi U(z)}{a}\right). \tag{2.7}$$

Solution of equation (2.7) with the boundary condition $U(\infty)=0$, $U(-\infty)=a$ is well-known in the theory of dislocation [101] and has the form

$$U(z)=\frac{a}{2}\left[1-\frac{2}{\pi}arctg\frac{(z-Z)}{\lambda_1}\right],\quad \lambda_1=\frac{4P_0^2a^2}{\pi V_0\sqrt{\varepsilon_c\varepsilon_a}}. \tag{2.8}$$

Here $Z$ is the coordinate of the middle of the charged lateral wall of the nucleus, $\lambda_1$ is its width. At $P_0\sim10^4$, $a\sim10^{-7}$ cm, $\varepsilon_c\sim10^3$, $\varepsilon_a\sim10$, $V_0\sim10^{-2}\div10^{-1}$ erg·cm$^{-2}$ we obtain $\lambda_1\sim10^{-6}\div10^{-7}$ cm, which justifies the possibility of use of the continual consideration in this case. The condition $\lambda_1\gg a$ of the small incline of the nucleus wall in relation to the nondisplaced boundary, makes it possible to place, when writing equations (2.5)–(2.8), the bound charge on the boundary into the plane of the nondisplaced boundary.

The energy of the charged wall of the nucleus consists of the energy of misalignment of the boundary with the minimum of the potential relief $\gamma(U)$ and the electrostatic energy of the bound charges in it.

The linear density of misalignment energy is

$$W_v=\int_{-\infty}^{\infty}(\gamma(U)-\gamma_0)dz=-\int_{-\infty}^{\infty}zd\gamma(U)=-\int_{-\infty}^{\infty}z\frac{d\gamma}{dU}\frac{dU}{dz}dz. \tag{2.9}$$

Substituting here $d\gamma/dU$, on the basis of (2.7) we get

$$W_v=-\frac{8P_0^2}{\sqrt{\varepsilon_c\varepsilon_a}}\int_{-\infty}^{\infty}\int_{-\infty}^{\infty}\frac{dU(z')}{dz'}\frac{dU}{dz}z\frac{dzdz'}{(z-z')}. \tag{2.10}$$

Adding to the integral in (2.10) the equivalent integral, where $z$ and $z'$ change their places, we obtain

$$W_v=\frac{4P_0^2a^2}{\sqrt{\varepsilon_c\varepsilon_a}}. \tag{2.11}$$

The linear density of the electrostatic energy of the charged wall of the nucleus

$$W_q=\frac{1}{2}\int\sigma(z)\varphi(z)dz \tag{2.12}$$

taking into account

$$\varphi(z)=\int\frac{2\sigma(z)}{\sqrt{\varepsilon_c\varepsilon_a}}\ln(z-z')dz' \tag{2.13}$$

is

$$W_q = \frac{4P_0^2 a^2}{\sqrt{\varepsilon_c \varepsilon_a}} \ln(\lambda_1/a). \tag{2.14}$$

The structure of the uncharged wall of the nucleus is determined by the equation similar to (2.7), in which the left hand part is substituted by the Laplace pressure:

$$\gamma_0 \frac{d^2U}{dy^2} = \frac{\pi V_0}{a} \sin\left(\frac{2\pi U(y)}{a}\right). \tag{2.15}$$

Integration of the equation, using the previous boundary conditions $U(\infty)=0$, $U(-\infty)=a$ (the period of the function $\gamma(U)$ in the direction $y$ is assumed also to be equal to $a$ to simplify consideration) gives [100]:

$$\frac{\gamma_0}{2}\left(\frac{dU}{dy}\right)^2 = \gamma(U) - \gamma_0 \tag{2.16}$$

and the equation for determination of the coordinate dependence $U(y)$:

$$\sqrt{\frac{\gamma_0}{2}} \int \frac{dU}{\sqrt{V_0} \sin(\pi U/a)} = y. \tag{2.17}$$

Hence, the distribution of the displacements in the uncharged wall of the nucleus is

$$U(y) = \frac{a}{2}\left[1 - \frac{2}{\pi} arctg\left(\exp(-\pi x/\lambda_2)\right)\right], \quad \lambda_2 = a\sqrt{\frac{\gamma_0}{2V_0}}. \tag{2.18}$$

The linear density of the energy of the uncharged wall of the nucleus, linked with the increase of the total surface of the domain wall transient into the adjacent $x$=const plane

$$W_\gamma = \frac{\gamma_0}{2} \int_{-\infty}^{\infty} \left(\frac{dU}{dy}\right)^2 dy = \int \gamma(U) dy = W_\upsilon \tag{2.19}$$

is equal to the linear density of the misalignment energy. Therefore, the total density of the energy of the uncharged wall of the nucleus is

$$W_\gamma + W_\upsilon = 2\int_{-\infty}^{\infty} \left(\gamma(U) - \gamma_0\right) dy = \frac{2a}{\pi}\sqrt{2\gamma_0 V_0}. \tag{2.20}$$

The width of the uncharged wall at $V_0 \sim 0{,}1\gamma_0$, $\gamma_0 \sim 1$ erg/cm$^2$ is ~2.5$a$. The energy of the charged wall, related to the unit of its length $\gamma_1=W_q+W_v$ is $\sim 4\cdot 10^{-7}$ erg/cm. The linear density of the energy of the uncharged wall $\gamma_2=W_\gamma+W_v$ at the same values of $\gamma_0$ and $V_0$ is $4\cdot 10^{-8}$ erg/cm, i.e. an order of magnitude lower than $\gamma_1$. The estimates made enable us to make assumptions about the contribution of surface tension into the energy and structure of the charged wall itself and, in particular, regard them as negligible, which justifies the application of equation (2.4) above in the determination of the structure of the charged wall.

The broadening of the lateral wall of the nucleus in comparison with the conventional domain wall decreases the height of the barrier (for a conventional wall its magnitude is $V_0$), surmounted by the wall when it moves in activation mode, by the multiplier $\exp(-\pi^2\lambda/a)$, which makes the motion of the nucleus wall almost insensitive to the magnitude of the given barrier. Under these conditions, the velocity of the lateral wall of the nucleus is controlled by the viscosity of a certain nature. Let us replace here static equation (2.7) and (2.15) by the equations of motion of the lateral walls of the nucleus by adding to these equations the terms $\eta\ \partial U/\partial t$, and $2P_0Ea$, where $\eta$ is the coefficient of viscosity of the domain wall (see the previous section), and $2P_0\,Ea$ is the pressure from the direction of the external field on the unit length of the lateral wall of the nucleus. Then, assuming for simplicity of considerations that the structure of the moving wall does not change in comparison with its static configuration, and taking into account the relationship $\partial U/\partial t=-\upsilon_1\partial U/\partial z=-\upsilon_2\partial U/\partial y$ for the velocities of the charged and uncharged lateral walls of the nucleus, we obtain the following respective equations

$$\upsilon_1=\frac{8P_0^3a^2}{\eta V_0\sqrt{\varepsilon_c\varepsilon_a}}E. \tag{2.21}$$

$$\upsilon_2=\frac{\sqrt{2}\sqrt{\gamma}P_0a}{\eta V_0}E. \tag{2.22}$$

Then

$$\upsilon_1/\upsilon_2=\lambda_1/\lambda_2=\frac{4\sqrt{2}P_0^2a^2}{\pi\sqrt{\varepsilon_c\varepsilon_a}\sqrt{\gamma V_0}}=\gamma_1/\gamma_2. \tag{2.23}$$

As shown below, the ratio of the dimensions of the critical nucleus is $z_{\max}/y_{\max}=2\sqrt{\gamma_1/\gamma_2}$. Since, in the ratio of the velocities there is a higher degree of the ratio $\gamma_1/\gamma_2$, then in the ratio of the dimensions of the nucleus and, as a rule $\gamma_1/\gamma_2\gg 1$, then in the process of motion (growth) of the nucleus we should expect that it will stretch even greater in the polar direction (Fig.6.5).

The effective mass of the lateral walls of the nucleus on the domain wall, related to the unit of their length, is determined as in (1.1) by the ratio

$$\gamma(\upsilon \ll c_0) - \gamma_{1,2} = a\int \frac{\rho}{2}\left(\frac{dU_{1,2}}{dt}\right)^2 dx =$$
$$= \frac{\rho a}{2}\upsilon_{1,2}^2 \int_{-\infty}^{\infty}\left(\frac{dU_{1,2}}{dt}\right)^2 dx = \frac{m_{1,2}^* \upsilon_{1,2}^2}{2}. \quad (2.24)$$

Substituting here $dU_{1,2}/dx$ from (2.8) and (2.18) we obtain

$$m_{1,2}^* \simeq \rho a^3 / \lambda_{1,2}, \quad (2.25)$$

which gives the following expressions for the charged and uncharged sections of the wall of the nucleus respectively

$$m_1^* = \frac{\pi m V_0 \sqrt{\varepsilon_c \varepsilon_a}}{4P_0^2 a^2}, \quad (2.26)$$

$$m_2^* = \frac{m\sqrt{2V_0}}{a\sqrt{\gamma_0}}, \quad m = \rho a^3. \quad (2.27)$$

### 6.3. Velocity of the lateral motion of a domain wall of a ferroelectric under the conditions of thermofluctuation formation and growth of nuclei of inverse domains

To determine the parameters of a critical nucleus on a domain wall, we write a functional corresponding to the total energy of the nucleus

$$\Pi = \oint \gamma(\varphi)\, dl - 2P_0 E a \oint dS. \quad (3.1)$$

Here $\gamma(\varphi)$ is the linear density of the energy of the lateral wall of a flat nucleus as a function of its orientation, the angle $\varphi$ is determined by the ratio $\mathrm{tg}\varphi = y'$, where $y = y(z)$ is the coordinate dependence of the curve describ-

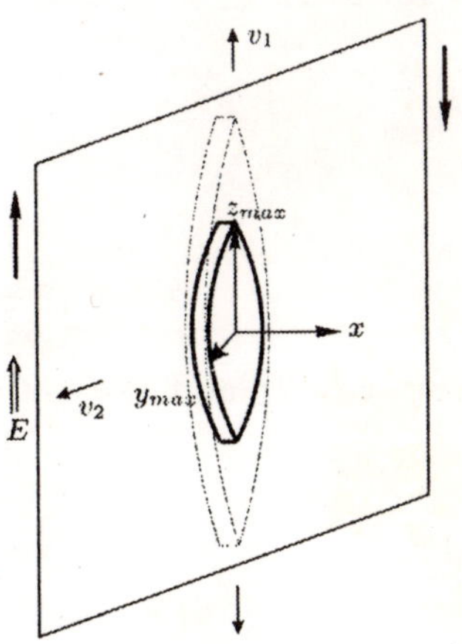


**Fig.6.5.** Critical nucleus on a domain wall. The broken line shows the change of the nucleus during its growth.

ing the boundary of the nucleus.

According to (2.11) and (2.14), the linear density of the energy of the charged lateral wall of the nucleus, parallel to the y axis, is proportional to the square of spontaneous polarization $P_0^2$. For a wall forming some angle with the lateral wall, the linear density of energy is determined evidently by replacing $P_0$ in (2.11) and (2.14) by the polarization component normal to the boundary of the nucleus and located in its plane. Taking this into account as well as the contribution of surface tension, the orientation dependence of the linear density of the energy of the lateral wall of the nucleus can be written in the form

$$\gamma(\varphi)=\gamma_1\cdot\sin^2\varphi+\gamma_2. \tag{3.2}$$

The functional (3.1), written taking into account the specific orientation of dependence $\gamma$ (3.2) has the following form in the Cartesian coordinates

$$\Pi=\oint\left(\gamma_1\frac{y'^2}{\left(1+y'^2\right)}+\gamma_2\right)\sqrt{1+y'^2}\,dz-2P_0Ea\oint y\,dz. \tag{3.3}$$

The Euler equation, corresponding to the extremum of the functional (3.3)

$$\frac{d\gamma}{dy'}\sqrt{1+y'^2}+\gamma(y')\frac{y'}{\sqrt{1+y'^2}}=-Lz+\text{const} \tag{3.4}$$

is the equation for determination of the equilibrium form of the critical nucleus. In (3.4) $L=2P_0Ea$; the corresponding constant is determined from the boundary conditions and is equal to zero in this case.

Equation (3.4) in parametric form is as follows:

$$\frac{d\gamma}{d\varphi}\cos\varphi+\gamma\sin\varphi=-Lz. \tag{3.5}$$

Its integration gives [157]

$$z=-\frac{1}{L}\left(\frac{d\gamma}{d\varphi}\cos\varphi+\gamma\sin\varphi\right). \tag{3.6}$$

Taking into account the relation $y'=\text{tg}\varphi=dy/dz$ and the ratio (3.6), the differential

$$dy=\frac{1}{L}\left(\gamma\sin\varphi+\frac{d^2\gamma}{d\varphi^2}\sin\varphi\right)d\varphi, \tag{3.7}$$

hence

$$y=\frac{1}{L}\left(\gamma\cos\varphi-\frac{d\gamma}{d\varphi}\sin\varphi\right). \tag{3.8}$$

Substituting in (3.6) and (3.8) $\gamma\equiv\gamma(\varphi)$ from (3.2), we obtain the equation for the boundary of nucleus in parametric form

$$\begin{cases} z=-\dfrac{1}{2P_0a}\left[\gamma_1\sin^3\varphi+\gamma_2\sin\varphi+2\gamma_1\sin\varphi\cos^2\varphi\right] \\ y=\dfrac{1}{2P_0a}\left[-\gamma_1\sin^2\varphi\cos\varphi+\gamma_2\cos\varphi\right]. \end{cases} \tag{3.9}$$

Analysis of the relations (3.9) shows that depending on the ratio between $\gamma_2$ and $\gamma_1$, the form of the critical nucleus can change qualitatively. In order to illustrate this, let us consider a section of the wall of the nucleus, resting on a unit base perpendicular with regard to the polar axis and forming angle $\varphi$ with it. The density of its energy is

$$\Pi=\frac{\gamma(\varphi)}{\sin\varphi}. \tag{3.10}$$

The minimality condition $\Pi(\dot{\Pi}=0)$ has the form

$$\dot{\gamma}\sin\varphi_0=\gamma\cos\varphi_0. \tag{3.11}$$

Substitution of $\gamma$ in the form of (3.2) into (3.11) makes it possible to find the optimum orientation of the considered wall of the nucleus from the ratio

$$\sin^2\varphi_0=\gamma_2/\gamma_1. \tag{3.12}$$

Equation (3.12) shows that an oval nucleus is stable only at $\gamma_2\geq\gamma_1$, at $\gamma_1>\gamma_2$ the oval form becomes unstable and the nucleus becomes lenticular with the angle $\varphi_0$ between the surfaces forming it in the area of their intersection (Fig. 6.6).

Condition (3.11) exactly corresponds to the conversion of coordinate $y$ to zero. Taking this into account, from the expression for $z$ (3.9) we obtain $z_{\max}=\sqrt{\gamma_1\gamma_2}/P_0Ea$. The maximum value of $y$ is equal to $y_{\max}=\gamma_2/2P_0Ea$.*Thus, their ratio $z_{\max}/y_{\max}=2\sqrt{\gamma_1/\gamma_2}$ is determined by the ratio of the linear densities of the energy of the charged and uncharged walls of the nucleus and in accordance with the actual relation between $\gamma_1$ and $\gamma_2$ indicates the elongation of the critical nucleus along the polar axis (Fig. 6.6).

To obtain the energy of the critical nucleus, let us write the

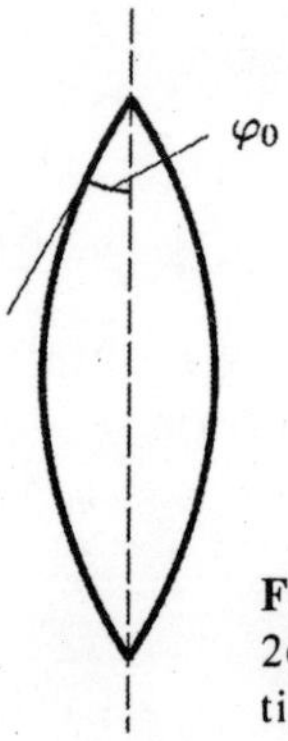


**Fig.6.6.** The form of a critical nucleus in the plane of the domain wall. $2\varphi_0$ is the angle between the forming surfaces in the area of the sharp tip of the lense.

functional (3.1) in the parametric form

$$\Pi=\oint\gamma(\varphi)\sqrt{\dot{z}^2+\dot{y}^2}\,d\varphi+\frac{L}{2}\oint(y\dot{z}-z\dot{y})\,d\varphi \qquad (3.13)$$

(here the dot indicates differentiation with respect to the angle $\varphi$). On the basis of (3.6) and (3.8)

$$\dot{z}=-\frac{1}{L}(\gamma+\ddot{\gamma})\cos\varphi,\quad \dot{y}=-\frac{1}{L}(\gamma+\ddot{\gamma})\sin\varphi. \qquad (3.14)$$

Then the energy of the critical nucleus is

$$\Pi^*=\frac{2}{L}\int_0^{\varphi_0}\gamma(\gamma+\ddot{\gamma})\,d\varphi. \qquad (3.15)$$

Substituting here $\gamma$ and $\ddot{\gamma}$, on the basis of (3.2) for the arbitrary ratio between $\gamma_1$ and $\gamma_2$ we obtain

$$\Pi^*=\frac{1}{P_0Ea}\sqrt{\frac{\gamma_2}{\gamma_1}\left(1-\frac{\gamma_2}{\gamma_1}\right)}\left\{\frac{\gamma_1^2}{8}+\frac{7}{4}\gamma_2\gamma_1\right\}+ \\ +\frac{1}{P_0Ea}\arcsin\sqrt{\frac{\gamma_2}{\gamma_1}}\left\{-\frac{\gamma_1^2}{8}+\gamma_1\gamma_2+\gamma_2^2\right\}. \qquad (3.16)$$

If $\gamma_1 \gg \gamma_2$

$$\Pi^*\simeq\frac{8}{3P_0Ea}\sqrt{\gamma_1\gamma_2^3}, \qquad (3.17)$$

in the inverse limiting case $\gamma_1 \ll \gamma_2$

$$\Pi^*\simeq\frac{\pi}{2P_0Ea}\gamma_2^2, \qquad (3.18)$$

at the moment of the alteration of the nucleus form from the oval to lenticular, when $1-\gamma_2/\gamma_1<<1$

$$\Pi^* \simeq \frac{\pi}{2P_0Ea}\left[\gamma_1\gamma_2+\gamma_2^2\right]. \tag{3.19}$$

When describing the velocity of the lateral motion of the domain wall in the area of action of the thermofluctuation mechanism of nuclei formation on the domain wall, three areas can be defined.

In the case of relatively weak fields when the time between the nucleation of two nuclei on the domain wall is long in comparison with the duration of spreading of a single nucleus over the entire area of the wall, the velocity of the lateral motion of the domain wall is determined by the time of formation of a single nucleus on the entire area $S$ of the wall and turns out to be as follows:

$$\upsilon = aNS, \tag{3.20}$$

where

$$N = \frac{\nu_0}{s^*}\exp\left(-\Pi^*/T\right) \tag{3.21}$$

is the average number of nuclei formed during the time unit per unit of the domain wall area. Here $\nu_0$ is the characteristic frequency factor, $s^*$ is the area of the critical nucleus, which in the most realistic situation of $\gamma_1 \gg \gamma_2$ is $s^* = 4\sqrt{\gamma_1\gamma_2^3}/3P_0^2E^2a^2$.

In the range of intermediate fields, when many nuclei develop simultaneously on the wall, the velocity of lateral motion of the domain wall is determined not only by the probability of formation of nuclei on it, but also by their growth rate. As shown in the previous section, the growth rate of different sections of the nucleus varies.

The velocity of lateral motion of the domain wall in the range of intermediate fields is

$$\upsilon = a\sqrt[3]{4\upsilon_1\upsilon_2 N}. \tag{3.22}$$

In the presence of strong fields, when the displacement of the wall to the adjacent plane takes place only as a result of the formation of the required number of critical nuclei on it

$$\upsilon = aNs^*. \tag{3.23}$$

The velocity of the lateral motion of the domain wall here, as in the case of weak fields, does not depend on $\upsilon_1$, $\upsilon_2$.

Substituting $N$ from (3.21), into (3.20), (3.22), (3.23), we have for all three velocity regimes the activation character of motion with the activation field

$$\overline{\delta} = \frac{8\sqrt{\gamma_1 \gamma_2^3}}{3P_0 aT} \tag{3.24}$$

in the case of weak and strong fields and field $\overline{\delta}/3$ in the case of intermediate fields. Substitution of the specific values $\gamma_1$ and $\gamma_2$ into (2.24) shows that value $\overline{\delta}$ decreases at $T \to T_c$. To obtain the law of decrease of $\overline{\delta}$ it is necessary to known the temperature dependence of barrier $V_0$. If the temperature dependence of $V_0$ is the same as that of the energy of the domain wall $\gamma$, then $\overline{\delta}$ proves to be $\sim\Delta T^{3/2}$.

Let us give another result obtained from the analysis of the dependence $\upsilon(E)$ in various velocity regimes. As shown in equations (3.20) and (3.23) for $\upsilon$ in the case of weak and strong fields, and the expression for $s^*$, the pre-exponential multiplier in the dependence of $\upsilon(E)$ is proportional to $E^2$ in the case of weak fields and independent of the field in the case of strong fields. The mentioned alteration of the pre-exponential multiplier in the expression $\upsilon(E)$ in the assumption that it doesn't depend on the field can be interpreted also as some increase of activation field $\overline{\delta}$ with the increase of the strength of the applied field $E$ that was experimentally observed. It should be noted that the result is obtained here while considering the nuclei with the thickness of the constant of the elementary cell and, consequently, does not require taking into account the multilayer nuclei, which was proposed in [153,154].

The consideration above was based on the approximation of an ideal defect-free material. In real crystals, as shown in experimental observation [158–160], in addition to the Peierls relief the influence of crystalline lattice defects has to be considered as well [161–163].

## 6.4. Influence of tunnelling of ferroactive particles and temperature on the mobility of domain boundaries

To determine the influence of tunnelling on the mobility of the boundaries in the regime of the thermofluctuation mechanism of motion, specific equations for $\gamma_1$ and $\gamma_2$ are substituted into equation (3.24) for the activation field. This yields

$$\overline{\delta} = 2\left(\gamma V_0 a^2\right)^{3/4} / \left(\varepsilon_c \varepsilon_a\right)^{1/4} T. \tag{4.1}$$

In the quasi-continual approximation, the dependence of the magnitude of the lattice energy barrier $V_0$ on the parameters of the domain wall has the form (1.10)

$$V_0 = 8\pi^4\gamma\left(\frac{\delta}{a}\right)^3 \exp\left(-\frac{\pi^2\delta}{a}\right). \tag{4.2}$$

Substituting (4.2) into (4.1) we obtain the following for the velocity of the lateral motion of the domain wall:

$$\upsilon = \upsilon_\infty \exp\left\{-W\exp\left\{-\frac{3\pi^2}{4}\frac{\delta}{a}\right\}\right\},$$

$$W = 2\left(\gamma a^2\right)^{3/4}\left(\varepsilon_c\varepsilon_a\right)^{-1/4} T^{-1}\left(8\pi^4\gamma\right)^{3/4}\left(\frac{\delta}{a}\right)^{9/4} E^{-1}. \tag{4.3}$$

The ratios (4.3) show that the velocity of the domain wall in the given regime very strongly depends on its width (the functional dependence is exponent in exponent). Therefore, even a relatively small increase of the width of the wall $\delta$ as a result of the tunnelling effect should result in a considerable increase of the velocity of the wall $v$.

Evidently, this increase can be one of the reasons for the increase of the mobility of the domain walls by six orders of magnitude at once observed in the experiments [78] when replacing deuterium by hydrogen in the structure of $KD_2PO_4$. In fact, at the above values of the constants $(\mathcal{J}+2A)_D \simeq 213$ K, $(\mathcal{J}+2A)_H \simeq 140$ K, $A_D \simeq 20$ K, $\Omega_D \simeq 0$ K, $\Omega_H \simeq 86$ K, $a \sim 10^{-7}$ cm at $\Delta T \sim 20$ K, where $\gamma_D \simeq 4.2\cdot 10^{-2}$ erg.cm$^{-2}$, $\gamma_H \simeq 3.5\cdot 10^{-2}$ erg.cm$^{-2}$ and $(\delta/a)_D \simeq 1{,}4$, $(\delta/a)_H \simeq 2$, $\varepsilon_c^H \sim 10^3$, $\varepsilon_c^D \sim 10^2$, $\varepsilon_a^H \sim 10$, $\varepsilon_a^D \sim 5$ at $E \sim 1$ V$\cdot$cm$^{-1}$ the value of the constant $W$ for the deuterated and undeuterated crystals is equal to respectively $W_D \simeq 5\cdot 10^5$, $W_H \simeq 7.7\cdot 10^5$.

Assuming hereinafter that the pre-exponential multiplier in (4.3) does not change at substitution $H \rightarrow D$, taking into account the obtained values of $W_H$ and $W_D$ for the ratio of the velocity of the domain boundaries in the deuterated and undeuterated crystals equidistant from $T_c$ by the value $\Delta T \sim 20$ K, we have $\upsilon_D/\upsilon_H \sim 10^{-5} \div 10^{-6}$.

In the regime of viscose motion of the domain walls with the linear dependence of their velocity on the strength of the external field

$$\upsilon = \frac{3}{2}\frac{\delta}{P_0\Gamma}E \tag{4.4}$$

the value of the latter velocity as seen from (4.4), is inversely proportional to the curvature of distribution of the order parameter in the boundary $P_0/\delta$. and since according to the results of section 3.6, the value of the latter decreases with increase of the tunnelling effect, then in the given case we have an increase of the velocity of the domain wall with increasing $\Omega$.

Substituting in (4.4) ratio $Z/\delta$ (6.11) from part 3.6 multiplied by $P_0$ instead of ratio $P_0/\delta$, we obtain

$$\upsilon = \frac{E}{P_0} \frac{\sqrt{A \cdot T_c} \cdot a}{(T - T_c)\left[1 - \Omega^2 / T_c^2\right]\Gamma}. \qquad (4.5)$$

According to (4.5), together with the decrease of a curvature of distribution of the order parameter in the boundary at approaching $T_c$ the velocity of the motion of the domain wall in the given external field $E$ also increases. A similar but considerably stronger dependence $\upsilon(T)$ follows from ratio (4.3).

## 6.5. 'Freezing' of the domain structure in the crystals of the $KH_2PO_4$(KDP) group

In any of the motion regimes (3.20), (3.22), (3.23) the velocity of the domain wall, controlled by the thermofluctuation surpassing of the Peierls-type barrier, considerably depends on temperature. Evidently, the latter can be detected not only by direct measurements but also as a result of indirect investigations, one of which is the study of the temperature dependence of dielectric permittivity.

On the basis of direct optical and indirect dielectric measurements, the anomalously strong dependence of the velocity of the domain wall on temperature at specific temperatures is found in the case of crystals of the $KH_2PO_4$ group. The specifics of the dielectric properties of the crystals of this group is the presence of the strongly distinguished domain contribution to the values of the dielectric constant measured along the polar axis, in the limits of the so-called 'plateau' region (region of almost constant values of $\varepsilon$), Fig.6.7. However, it is evident that even more distinguishing special features of this type of crystals is the rapid disappearance of the mentioned domain contribution to the values of $\varepsilon$ at some temperature $T_f$ with the simultaneous increase of the values of the dielectric loss angle tangent tg$\delta$ at these temperatures. This phenomenon was called 'freezing' of the domain structure [164–168].

The general form of the dependences $\varepsilon(T)$ and tg$\delta(T)$ is similar for all isomorphous KDP crystals (Fig.6.7), although quantitatively

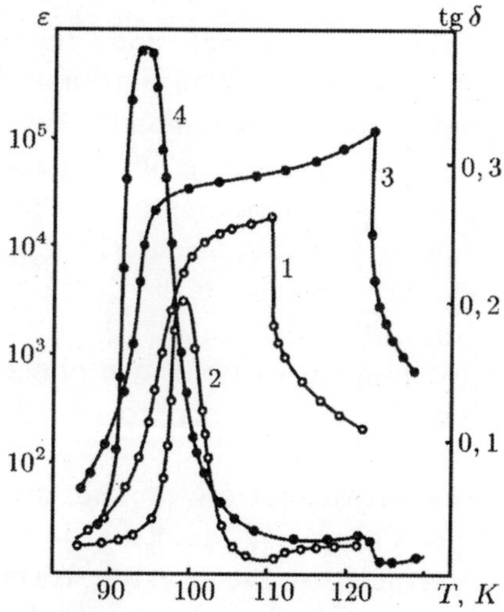


**Fig.6.7.** Temperature dependences of $\varepsilon$ (1,3) and tg$\delta$ (2,4) for crystals of $RbH_2AsO_4$ (1,2) and $KH_2PO_4$(3,4). $E_{\sim}$=1 V·cm$^{-1}$, $f$=1 kHz.

the arsenates show lower values of $\varepsilon(\sim 10^3)$ in the 'plateau' region in comparison with phosphates (for the phosphates $\sim 10^5$), they also have a lower general level of the dielectric losses and a narrower region of the 'plateau' itself [169]. The values of the mentioned characteristics depend strongly on the presence of specific influences on the domain structure and measurement conditions, in particular, on the concentration of structural defects [170] and the amplitude $E_{\sim}$ of the measuring field [171,172]. Unlike other phosphates, in the $CsH_2PO_4$ crystal with the structure of the ferrophase differing from those of the previously mentioned crystals, in the field with $E_{\sim}$ ~1 V·cm$^{-1}$ the width of the 'plateau' region is very small (of the order of several degrees, whereas for other crystals it can be equal to several tens of degrees) [173–178].

Deuteration results in a large decrease of the value of $\varepsilon$ on the 'plateau' (approximately by two orders of magnitude) and in a smoother (in comparison with the undeuterated crystals) decrease of the values of $\varepsilon$ and a less distinctive maximum of tg$\delta$ in the vicinity of $T_f$ [179–181].

The first considerations regarding the nature of 'freezing' of the domain structure in the crystals of the $KH_2PO_4$ group were published by Barkla and Finlayson who experimentally detected this phenomenon [164]. However, their assumption regarding the actual destruction of the crystal at temperature $T_f$ has not been confirmed.

The 'freezing' phenomenon has been discussed most thoroughly in [182–184], where an assumption was made regarding its elastic nature. The symmetry alteration at phase transition to the polar state from the tetragonal to the orthorhombic in crystals of the KDP group is accompanied by the occurrence here of spontaneous shear strain $x_y = \epsilon_0$ in the plane perpendicular to the polar axis, which allows the existence of four possible orientations of the elementary cells [2]. In compliance with the above the domain structure in the KDP-type crystals should represent 180° blocks of $x$- and $y$-domains, separated by domain walls, whose planes are parallel to the polar direction (axis $z$) and normal to the shear plane (Fig. 6.8).

The real conditions of formation of the domain structure, associated with the intergrowth of blocks growing towards each other from opposite surfaces of the crystal, form needle-shaped domains elongated along one of the two tetragonal axes $x$ or $y$, with the width $d$ along the other axis, constant in the specimen at the zero value of the external electric field or mechanical stress. The tips of the needles are sometimes wedge-shaped, but in most cases they are rounded. At that the domains grow entirely through the crystal plate in the polar direction [182].

At the tips of the domains where the domain boundary leaves the planes (100) and (010), in accordance with the symmetry of the low temperature phase, there is a lot of edge twinning dislocations discussed in section 4.2. The mobility of these dislocations in the natural Peierls relief under the condition of the intergrowth of the domain tip is the factor, determining, according to [182], the general ability of the domain boundaries to displace in the process of repolarization and consequently their contribution to $\varepsilon$, since [182] assumes strong correlation between the motion of the tip of the domain and the lateral motion of domain boundaries.

The viewpoint described in [182–184] has its own flaws. The point is that, like the motion of the domain boundaries in the lattice

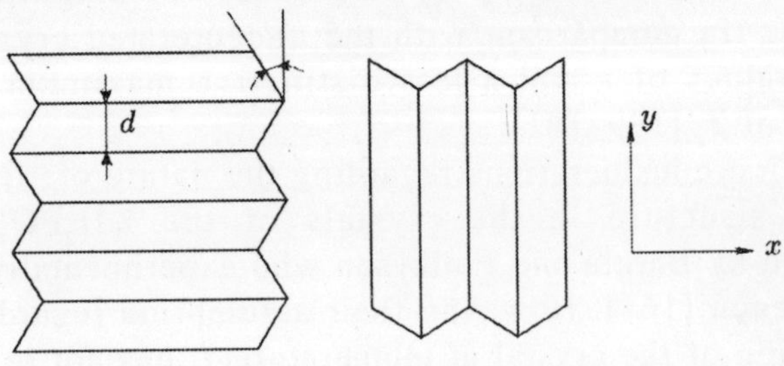


**Fig.6.8.** Domain structure in crystals of the $KH_2PO_4$ group.

relief, the motion of twinning dislocations in the Peierls relief requires a critical mechanical stress or taking into account the fact that the crystal under consideration is a ferroelectric – ferroelastic – a critical electric field. As with the motion of the unbent domain boundaries, the strength of the latter is

$$E_{cr} \sim \exp\left\{-\frac{4\pi\zeta}{b}\right\}, \tag{5.1}$$

where $\zeta$ is the half width of the dislocation, and $b$ is its Burger's vector. Value $\zeta$ in (5.1) is equal to $\zeta=a/2(1-\nu)$, where $\nu=\lambda/2(\lambda+\mu)$ is the Poisson coefficient and $b=2a\epsilon_0$. In the index of the exponent of expression (5.1) the only value which depends on temperature could be the length of the vector $b$ due to the possible temperature dependence of spontaneous strain $\epsilon_0$. However, as shown by the experimental investigations of the spontaneous polarisation linked linearly with $\epsilon_0$ [2], in the 'freezing' temperature range of the domain structure, these values are almost independent of temperature.

The above mentioned weak dependence of the critical field on temperature contradicts experiments that studied the influence of the amplitude of the measuring field $E_{\sim}$ on the position of $T_f$ [168,171,172] where it was shown that although $T_f$ is shifted in the fields with a small amplitude, to obtain the given shift of the order of several degrees it is sometimes necessary to increase $E_{\sim}$ tens of times (Figs.6.9 and 6.10).

It should also be noted that since the values of the elastic constants for the crystals of isomorphic KDP are close in magnitude to each other in the framework of the model [182], it is difficult to understand a large difference in the values of $T_f$, for example, in the crystals of $CsH_2AsO_4$ ($T_f \simeq 140$ K), $CsH_2PO_4$ ($T_f \simeq 150$ K) [169,173] in comparison with other crystals of this group: $KH_2PO_4$, $RbH_2PO_4$, $RbH_2AsO_4$ ($T_f \simeq 95 \div 97$ K).

To explain the nature of the domain structure 'freezing', together with the already mentioned experimental data, let us use the frequency dependences of the components of dielectric permittivity in crystals of the $KH_2PO_4$ group. According to the results of [167], the frequency shift of $T_f$ observed here is very special: a large displacement of the maximum of dielectric losses towards high temperatures, which is usually regarded as an indication of relaxation losses, was noted in $KH_2PO_4$ only at $f \sim 10^7 \div 10^8$ Hz. With the change of the frequency of the measuring field in the range up to $10^7$ Hz, the position of the maximum mentioned above remains

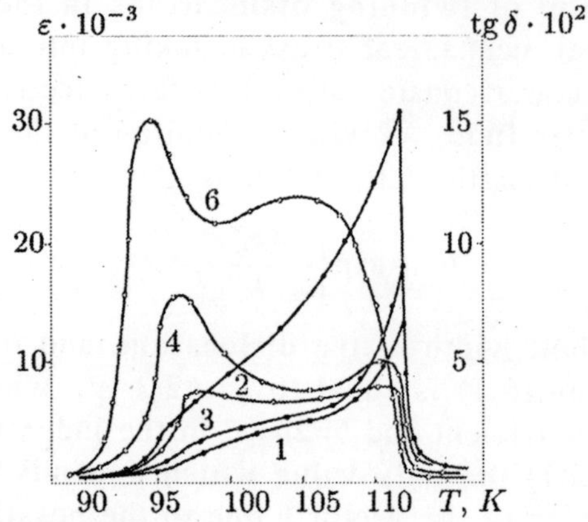


**Fig.6.9.** Temperature dependences of ε – 1, 3, 5 and tgδ(T) – 2,4,6 for a $RbH_2AsO_4$ crystal at several amplitudes of the measuring field: 1,2 – $E_{\sim}$=1; 3,4 – 5; 5,6 – 25 V·cm⁻¹.

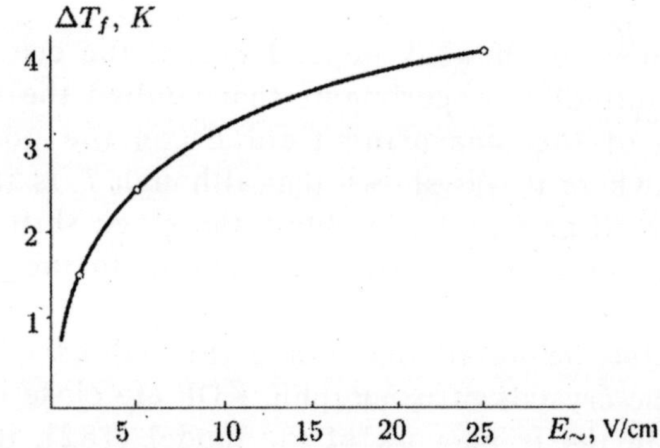


**Fig.6.10.** Dependence of the shift of the 'freezing' temperature of the domain structure ($T_f$=$T_{max\ tg\delta}$) for a $RbH_2AsO_4$ crystal on amplitude $E_{\sim}$ of the measuring field.

unchanged on the temperature scale.

The absence of monotonicity in the dependence $T_f(f)$ indicates the strong temperature dependence of the velocity of lateral motion of the domain wall, and in particular, the strong temperature dependence of the energy of the critical nucleus $\Pi^*$ on the domain wall, which increase sharply in the vicinity of $T_f$.

In fact, the location of the maximum of tg$\delta$ for relaxation losses (see chapter 7) is determined as usual by the condition $\omega\tau$ = const. According to this condition, the absence of the frequency shift of the maximum of tg$\delta$ with the variation of $f$ from $10^1$ to $10^7$ Hz indicates that in the vicinity of $T_f$ in the temperature range ~1 K

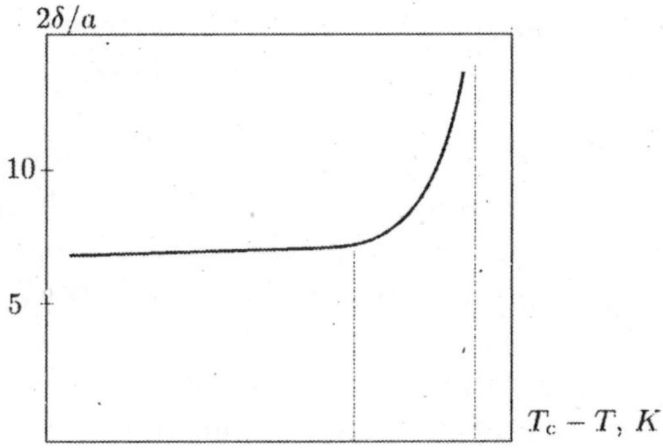


**Fig.6.11.** Temperature dependence of width of the domain wall in KDP [187].

the domain structure relaxation time changes by six orders of magnitude at once, whereas at $T>T_f$ in the temperature range greater than ten degrees (shift of $T_f$ at the variation of $f$ from $10^7$ to $10^8$ [167]) it changes by only an order of magnitude. In the vicinity of $T_f$, none of the parameters, determining $\tau$, except for $\Pi^*$, shows any irregular temperature dependence and, therefore, we should look for the reason of non-monotonicity in the dependence $\tau(T)$ and consequently, the explanation of the phenomenon of the domain structure 'freezing' in the critical dependence $\Pi^*(T)$ in the vicinity of $T_f$.

As it was shown by calculations in chapter 3, the reason for the occurrence of the phenomenon of 'freezing' here is most probably the rapid increase of the value of the lattice energy barrier $V_0$ (the energy of the critical nucleus $\Pi^* \sim V_0^{3/4}$), surpassed by the wall during lateral motion. As shown previously, in such a case the width of the domain wall, which decreases with the decrease of temperature, becomes comparable with the lattice spacing [87–89,185,186], which makes the wall especially sensitive to its position in the lattice potential relief. Evidently, the direct proof of this are the results of the study of the domain wall temperature dependence in the KDP crystal on the basis of analysis of the spectra of scattering of $x$-rays [187] (Fig.6.11), where it was found that the width of the domain wall receives a constant although a higher value below $T_f$. The interpretation of 'freezing' of the domain structure, proposed here, is also supported by the anomalous proximity to $T_c$ of the 'freezing' temperature in the quasi-one-dimensional ferroelectric $CsH_2PO_4$ [188] which possesses the minimum value of the constant $A$ amongst other crystals in its group. It should be noted that this phenomenon was predicted by equation (6.7) in chapter 3.

The knowledge of the temperature dependence $V_0(T)$ and general expressions (3.20)–(3.23) which determine the velocity of the lateral motion of the domain wall under the conditions of thermofluctuation formation and subsequent growth of the nuclei of inverse domains on the lateral surface of the domain wall makes it possible to determine directly the value of $T_f$ and also its dependence on the amplitude $E_\sim$ of the measuring field.

To determine the latter, let us write the expression for the frequency dependence of dielectric permittivity and use the Debye equation for this purpose

$$\varepsilon(\omega)=\varepsilon_\infty+\frac{\varepsilon_0-\varepsilon_\infty}{1-i\omega\tau}. \qquad (5.2)$$

Here $\varepsilon_0$ is the static dielectric permittivity, determined by the displacement of the domain boundaries, $\tau$ is the relaxation time of the domain structure. The latter can be found as the ratio

$$\tau=\frac{\overline{U}}{v}\simeq\frac{\overline{U}}{v_\infty}\cdot\exp\left(\overline{\delta}/E_\sim\right) \qquad (5.3)$$

where the average displacement of the domain walls $\overline{U}$, corresponding to static dielectric permittivity $\varepsilon_0$, is

$$\overline{U}=\varepsilon_0 E_\sim d/8\pi P_0. \qquad (5.4)$$

The location of the maximum of the tangent of the angle of dielectric losses from (5.2) is determined by the condition

$$\omega\tau_{max}=\sqrt{\varepsilon/\varepsilon_0}, \qquad (5.5)$$

whence the activation field, corresponding to the temperature of maximum tg$\delta$ is

$$\overline{\delta}_{\max}=E_\sim\ln\left(\frac{8\pi P_0 v_\infty}{\omega\sqrt{\varepsilon\varepsilon_0}dE_\sim}\right). \qquad (5.6)$$

On the basis of the ratio (5.6) and the expression for $\overline{\delta}$ (3.1), the value of the lattice barrier, determining the location of the maximum losses, is

$$V_{0\max}^{3/4}=V_0^{3/4}\left(T=T_f\right)=$$

$$=\frac{\left(\varepsilon_a\varepsilon_c\right)^{1/4}T_f}{16}\left(\frac{\pi}{2a\sqrt{2\gamma}}\right)^{3/2}\ln\left(\frac{8\pi P_0 v_\infty}{\omega\sqrt{\varepsilon\varepsilon_0}dE_\sim}\right)E_\sim=A\cdot E_\sim. \qquad (5.7)$$

According to the direct calculations in section 3.5, the temperature dependence of $V_0$ for the crystals of the investigated type can be characterized as follows. In the 'plateau' region the value $V_0$ is so small that here in the fields with $\omega/2\pi<10^7$ Hz, we evidently measure the dielectric permittivity $\varepsilon$ controlled by quasi-elastic displacements of the boundaries. With the decrease of temperature the value of $V_0$ reaches its minimum at the rearrangement temperature in the domain boundary $T_0$ and, then, with the further decrease of temperature it rapidly increases in accordance with the law, which can be approximated by the power dependence: $V_0=c\cdot\Delta T^n+V_0(T_0)$, where $\Delta T=T-T_0$, with index $n>1$. On the basis of the temperature dependence of $V_0$ and ratio (5.7), the shift of the 'freezing' temperature with increasing amplitude of the measuring field takes place in the direction of low temperatures (with the increase of $E_\sim$ decreases the energy of the critical nucleus) and at $V_0(T_0)\simeq 0$ it is governed by the law

$$\Delta T_f \simeq \left(\frac{A^{4/3}}{C}\right)^{1/n} E_\sim^{\alpha}, \quad \alpha=\frac{4}{3}\cdot\frac{1}{n}. \tag{5.8}$$

Taking into account the real value $n>1$, from ratio (5.8) we have $\alpha<1$ that, in fact, corresponds to the experimental results presented in Figs.6.9 and 6.10 which show that with the increase of the amplitude of the measuring field the experimentally measured shift of the 'freezing' temperature of the domain structure gradually slows down.

Let us estimate the coefficient of proportionality between $\Delta T_f$ and $E_\sim$ in (5.8). At $\varepsilon_c\sim 10^3$, $\varepsilon_a\sim 10$, $T_f\sim 10^{-14}$, $a\sim 10^{-7}$, $\gamma\sim 0.1$, $P_0\sim 10^4$, $\omega\sim 10^3$, $E_\sim\sim 10^{-2}$, $d\sim 10^4$ CGSE units we have $A\sim 10^{-1}$. The value of $C$, according to calculations in section 3.5, is estimated as $\sim 10^{-2}\div 10^{-3}$. Thus, the coefficient of proportionality between $\Delta T_f$ and $E_\sim$ is $\sim 10^{-2}\div 10^{-1}$ which at $n\sim 2$, $a\sim 2/3$ gives the shift $\Delta T_f$ of several degrees while the field $E_\sim$ increases from 1 to 25 V·cm$^{-1}$. This is also in good agreement with the experiments, where the shift of $T_f$ of several degrees in weak fields $E_\sim<10^2$ V·cm$^{-1}$ is typical of all crystals of the KDP group. For example, in the $RbH_2AsO_4$ crystal (Figs.6.9 and 6.10) the shift of $T_f$ with the increase of $E_\sim$ from 1 to 25 V·cm$^{-1}$ reaches 5 degrees.

## Chapter 7

# Natural and forced dynamics of boundaries in crystals of ferroelectrics and ferroelastics

### 7.1. BENDING VIBRATIONS OF 180° DOMAIN BOUNDARIES OF DEFECT-FREE FERROELECTRICS

This final chapter is devoted to the study of the dynamic aspects of the domain boundaries motion in ferroelectrics and ferroelastics. Let us start with the study of the dynamics of domain wall bending vibrations in pure defect-free ferroelectrics with 180° domain structure.

As it was already proved in Chapters 4 and 5 while studying the problems of interaction of the domain boundaries with defects and the problems of stability of the profile or orientation of the domain walls, the deviation of the latter from the polar axis in the ferroelectrics or from the direction of spontaneous shear in the ferroelastics increases the energy of the system due to the formation of the long-range electrical or elastic fields. This is equivalent to the situation, in which the bent domain walls are subjected to the restoring force, which in the approximation of small displacements of the wall is linear in relation to value $U$ of displacement of the wall and, consequently, may be regarded as quasi-elastic one.

The presence, in addition to this force, of the inertial properties of domain boundaries, discussed in Section 6.1, and also in this and subsequent sections, will lead to the formation of suitable conditions for bending vibrations of domain walls [189, 190].

A special feature of this motion of the domain boundaries in ferroelectrics, as of any other motion connected with the formation of long-range electric fields under the displacement of domain walls, is the involvement into the motion by way of the piezoelectric effect of the elastic medium, surrounding the boundary [191–195]. As shown later, in comparison with the situation when the piezoelectric effect is not taken into account, this results only in a small addition to the coefficient of the quasi-elastic force acting on the displaced boundary but has a drastic effect on its effective mass and, consequently, on the dynamics of the domain walls.

Taking the above into account, to obtain the law of dispersion of the domain boundaries bending vibrations in a pure defect-free ferroelectric we will use the equation of motion of the domain wall

$$m\ddot{U} = 2(\mathbf{P}_0, \mathbf{E})_{U=0}, \tag{1.1}$$

supplemented by the equation of motion of the elastic medium and the electrostatic equation

$$\rho \ddot{u}_{ij} = \frac{\partial \sigma_{ij}}{\partial x_j}, \quad \frac{\partial D_i}{\partial x_i} = 0, \tag{1.2}$$

where $m$ is the local effective mass of the domain wall, $D_i$ is the vector of electrostatic induction.

Writing equations for the components of the tensor of elastic stresses $\sigma_{ij}$ and vector $D_i$ in a crystal with a piezoelectric effect [196]

$$\begin{cases} \sigma_{ij} = c_{ijkl} u_{kl} + \beta_{kij} E_k, \\ D_i = \varepsilon_{ij} E_{ij} + 4\pi P_{0i} - 4\pi \beta_{ijk} u_{jk}. \end{cases} \tag{1.3}$$

where, as previously, $c_{ijkl}$, $\varepsilon_{ij}$ are the tensors of the elastic moduli and dielectric permittivity of the monodomain crystal, respectively, and $\beta_{ijk}$ is the tensor of piezoelectric coefficients, for the case of 180° domain wall, located in the nondisplaced position in plane $zy$ and the polar direction coinciding with axis $z$, taking into account the distribution of polarization in the crystal, containing the domain wall

$$P_{03}(z) = -P_0\left[1 - 2\Theta(x - U)\right], \tag{1.4}$$

the expression for the strain tensor

$$u_{jk} = \frac{1}{2}\left(\frac{\partial u_j}{\partial x_k} + \frac{\partial u_k}{\partial x_j}\right), \tag{1.5}$$

and the link to the strength of the electric field with potential $E_i = -\partial\varphi/\partial x_i$, we rewrite the set of equations (1.1)–(1.2) in the form

$$\begin{cases} \rho \ddot{u}_i = c_{ijkl}\dfrac{\partial^2 u_k}{\partial x_l \partial x_j} - \beta_{ijk}\dfrac{\partial^2 \varphi}{\partial x_k \partial x_j}, \\ -\varepsilon_{ij}\dfrac{\partial^2 \varphi}{\partial x_i \partial x_j} - 4\pi\beta_{ijk}\dfrac{\partial^2 u_j}{\partial x_k \partial x_i} = 8\pi P_0 \delta(x)\dfrac{\partial U}{\partial z}, \\ m\ddot{U} = 2P_0 \left.\dfrac{\partial \varphi}{\partial z}\right|_{x=0}. \end{cases} \tag{1.6}$$

Let us find an expression for the electric field $(-\partial\varphi/\partial x)_{x=0}$, accompanying bending displacement of the domain walis. To derive this equation, let us use the first two equations in (1.6). For an elastic–isotropic material, where

$$c_{ijkl} = \lambda\delta_{ij}\delta_{kl} + \mu\left(\delta_{ik}\delta_{jl} + \delta_{il}\delta_{jk}\right), \tag{1.7}$$

the equation of dynamics of the elastic medium has the following form

$$\rho\ddot{u}_i = (\lambda+\mu)\frac{\partial^2 u_l}{\partial x_l \partial x_i} + \mu\frac{\partial^2 u_i}{\partial x_k^2} - \beta_{kij}\frac{\partial^2 \varphi}{\partial x_k \partial x_j}. \tag{1.8}$$

Writing (1.8) in the vector form and using the operator *div* for the both parts of the equation, we obtain

$$\mathrm{div}\,\mathbf{u} = -\frac{\partial\Lambda_i/\partial x_i}{\rho\left(c_l^2\tilde{k}^2 - \omega^2\right)}, \quad \tilde{k}^2 = k_x^2 + k_y^2 + k_z^2, \tag{1.9}$$

where $\Lambda_i = \beta_{kij}\partial^2\varphi/\partial x_k\partial x_j$, and $c_l = \sqrt{(\lambda+2\mu)/\rho}$ is the velocity of the longitudinal sound wave. Taking this into account, instead of (1.8) we obtain

$$\rho\ddot{u}_i = -\frac{(\lambda+\mu)\partial^2\Lambda_l/\partial x_l\partial x_i}{\rho\left(c_l^2\tilde{k}^2 - \omega^2\right)} + \frac{\partial^2 u_i}{\partial x_k^2} - \Lambda_i. \tag{1.10}$$

When writing (1.9) and (1.10) we consider the wave propagating along the wall with the wave vector $\mathbf{k} = (k_y, k_z)$ and use the Fourier expansion for the vector of elastic displacement of the medium

$$u_i = \int u_i^{k_x} \cdot e^{i(wt - \mathbf{k}\boldsymbol{\rho})} \cdot e^{-ik_x x}\frac{dk_x}{2\pi}, \quad \boldsymbol{\rho} = (y, z). \tag{1.11}$$

Substituting (1.11) into (1.10), we obtain the ratio between the Fourier coefficients $u_i^{k_x}$ and $\varphi^{k_x}$:

$$u_i^{k_x} = \frac{1}{\rho\left(c_t^2\tilde{k}^2 - \omega^2\right)}\left[\beta_{kij}k_k k_j - \frac{\left(c_l^2 - c_t^2\right)}{\left(c_l^2\tilde{k}^2 - \omega^2\right)}\beta_{klj}k_k k_j k_l k_i\right]\varphi^{k_x},$$
$$c_t = \sqrt{\mu/\rho}. \tag{1.12}$$

Similarly, from the electrostatic equation we have

$$\varepsilon_{ij}k_i k_j \varphi^{k_x} + 4\pi\beta_{ijk}k_k k_i u_j^{k_x} = 8\pi P_0 i k_z U. \tag{1.13}$$

Substituting expression (1.12) into (1.13), instead of $u_j^{k_z}$, we find the expression for $\varphi^{k_x}$ from which

$$-\left.\frac{\partial\varphi}{\partial z}\right|_{x=0} =$$

$$= \int_{-\infty}^{\infty} \frac{-8\pi P_0 k_z^2 U \dfrac{dk_x}{2\pi}}{\left\{\varepsilon_{ij}k_i k_j + \dfrac{4\pi\beta_{imk}k_k k_i}{\rho\left(c_t^2\tilde{k}^2 - \omega^2\right)}\left[\beta_{pmj}k_p k_j - \dfrac{\left(c_l^2 - c_t^2\right)}{\left(c_l^2\tilde{k}^2 - \omega^2\right)}\cdot\beta_{plj}k_p k_j k_l k_m\right]\right\}}. \tag{1.14}$$

Substitution of (1.14) into the equation of the domain boundary motion (1.1) makes it possible to determine the spectrum of bending vibrations of the 180° domain walls in ferroelectric crystals of an arbitrary symmetry. As an example, let us consider a case of bending vibrations of 180° domain walls for a tetragonal polar phase. Here

$$\varepsilon_{ij} = \begin{pmatrix} \varepsilon_a & 0 & 0 \\ 0 & \varepsilon_a & 0 \\ 0 & 0 & \varepsilon_c \end{pmatrix} \tag{1.15}$$

and the matrix of the piezoelectric moduli $\beta_{imk} = \beta_{ikm}$ has the following non-zero coefficients: $\beta_{333} \equiv \beta_3$, $\beta_{322} = \beta_{311} \equiv \beta_2$, $\beta_{223} = \beta_{131} \equiv \beta_1$ [196]. In this case, the expression for the field (1.14) has the following form [194]:

$$-\left(\frac{\partial\varphi}{\partial z}\right)\Bigg|_{x=0}=\int_{-\infty}^{\infty}\frac{dk_x}{2\pi}\left(-8\pi P_0k_z^2U\right)\Big[\varepsilon_c k_z^2+\varepsilon_a\left(k_x^2+k_y^2\right)+$$
$$+\frac{4\pi}{\rho}\frac{k_z^2\left\{\beta_3^2k_z^2+\left[\left(\beta_2+\beta_1\right)^2+2\beta_1\beta_3\right]\left(k_y^2+k_x^2\right)\right\}+\beta_1\left(k_y^2+k_x^2\right)}{\left(c_t^2\tilde{k}^2-\omega^2\right)}-$$
$$-\frac{4\pi}{\rho}\frac{\left(c_l^2-c_t^2\right)\left[k_z^2\left[\beta_3k_z^2+\left(\beta_2+2\beta_1\right)\left(k_y^2+k_x^2\right)\right]\right]}{\left(c_t^2\tilde{k}^2-\omega^2\right)\left(c_l^2\tilde{k}^2-\omega^2\right)}\Bigg]^{-1}. \tag{1.16}$$

Expanding the integrand in (1.16) into a series in respect of $\omega^2$ with the accuracy to the terms of the first order in the approximation of smallness of $4\pi\beta^2/\rho\varepsilon\cdot c^2 << 1$, in particular, for $\varepsilon_a \simeq \varepsilon_c \equiv \varepsilon$, $\beta_3 \gg \beta_1$, $\beta_2$ and $k_z \gg k_y$, we obtain

$$-\frac{\partial\varphi}{\partial z}\bigg|_{x=0}\cdot P_0\simeq\frac{-4\pi P_0k_z^2U}{\varepsilon_a\sqrt{\frac{\varepsilon_c}{\varepsilon_a}k_z^2+k_y^2}}+\frac{4\pi\beta_3^2}{\rho\varepsilon c_l^4}\cdot\frac{5\pi P_0^2k_z^6U}{4\varepsilon\left(k_z^2+k_y^2\right)^{7/2}}\omega^2. \tag{1.17}$$

Substitution of (1.17) into the equation of motion of the domain wall results in the following equation determining the law of dispersion of the domain boundaries bending vibrations in ferroelectrics

$$\left(m+\frac{5\pi P_0^2\beta_3^2k_z^6}{\rho\varepsilon^2c_l^4\left(k_z^2+k_y^2\right)^{7/2}}\right)\omega^2=\frac{4\pi P_0^2k_z^2}{\varepsilon_a\sqrt{\frac{\varepsilon_c}{\varepsilon_a}k_z^2+k_y^2}}. \tag{1.18}$$

The expression in the round brackets in front of $\omega^2$ can be interpreted as the renormalised effective mass of the domain wall containing the non-local term $m^* \sim 1/k$, due to involvement in the motion of the entire layer of the material with the thickness equal to $1/k$ as a result of the piezoelectric effect. This layer surrounds the boundary and starts to move with the motion of the domain wall.

For comparison of values $m^*$ and $m$ let us present the maximum value $m^*$ for the given modulus $k$ for the case of the wave propagating along the polar direction in the form

$$m^*=\frac{4\pi\beta^2}{\rho\varepsilon c_l^2}\frac{5\pi P_0^2}{4\varepsilon c_l^2}\frac{1}{k}\simeq\left(\frac{4\pi\beta^2}{\rho\varepsilon c_l^2}\right)\frac{\gamma}{c_l^2}\frac{1}{k\delta}=\left(\frac{4\pi\beta^2}{\rho\varepsilon c_l^2}\right)m\frac{1}{k\delta}, \tag{1.19}$$

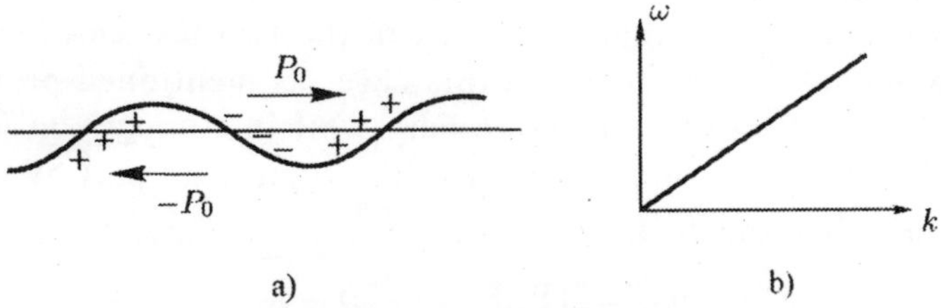


**Fig.7.1.** (*a*) – Formation of bound charges on the domain wall of a ferroelectric during its deflection from the polar direction. (*b*) – The linear dependence of the frequency of bending vibrations of the 180° domain boundaries on the wave vector taking into account the piezoelectric 'swinging' of the surrounding material.

where $\gamma$ is the surface density of energy of the domain wall, $\delta$ is the thickness of the domain wall. When writing (1.19) the expression for $\gamma$ (1.22) from Chapter 2 and for the local mass of the domain wall (1.6) from Chapter 6 were used.

Expression (1.19) shows that the value $m^*$ determines the law of dispersion of vibrations up to $k^* = 1/\delta\left(4\pi\beta^2/\rho\varepsilon c_l^2\right)$, i.e. taking into account the used approximation $\left(4\pi\beta^2/\rho\varepsilon c_l^2\right) \ll 1$ up to the value $k^* \ll k_{max} = 1/\delta$, where $k_{max}$ is the limiting value of the wave vector **k** determined by the limit of applicability of the approximation of the geometrical boundary. For the other orientations of **k** the value of $k^*$ is evidently lower.

The right-hand part of equation (1.18) with the accuracy to multiplier $1/k^2$ represents the effective rigidity relative to bending displacements of the domain wall. As it can be seen in the adopted approximation $(4\pi\beta^2/\rho\varepsilon c^2) \ll 1$ the contribution to it as a result of piezoelectric interactions is negligible and is completely determined by the electrostatic interaction of the charges on the bent boundary.

Analysing the law of dispersion of domain boundary vibrations under consideration, from equation (1.18) we can easily see that in the region of low $k$, where $m^* > m$, we have $\omega \sim k$ and, consequently, the velocity of propagation of the corresponding waves does not depend here on the wave vector.

## 7.2. BENDING VIBRATIONS OF DOMAIN BOUNDARIES OF DEFECT-FREE FERROELASTICS, FERROELECTRIC–FERROELASTICS AND 90° DOMAIN BOUNDARIES OF FERROELECTRICS

In pure ferroelastics, ferroelectrics–ferroelastics, and also in the case of 90° domain boundaries, for example, in ferroelectrics with a perovskite structure, the domains separated by them differ not only in spontaneous polarization, as in the two last cases, but also in spontaneous deformation. In this case, as mentioned previously, the equation of motion of the domain wall in comparison with the pure ferroelectric is supplemented in the right-hand part by the term $2\sigma_{ik}u^s_{ik}$ and has the form

$$m\ddot{U} = 2(\mathbf{P}_0,\mathbf{E})_{U=0} + 2\sigma_{ik}u^s_{ik}. \tag{2.1}$$

The presence of the direct elastic interaction in this problem results in the fact that its contribution to $\sigma_{ik}$ exceeds the contribution of the piezoelectric effect and, consequently, this contribution and not the piezoelectric effect sets the medium, surrounding the domain wall into motion, i.e. determines the effective mass of the domain wall. Taking this into account, the piezoelectric effect in the given problem can be ignored that greatly simplifies the calculation of the electric field. To determine the latter in this case the conventional equation of electrostatics can be used

$$-\varepsilon_{ij}\frac{\partial^2\varphi}{\partial x_i\partial x_j} = 8\pi P_0\delta(x)\frac{\partial U}{\partial z}. \tag{2.2}$$

To calculate the ratio of $\sigma_{ik}$ with the displacement of the wall $U$ we can use here the Beltrami dynamic equation (2.15) from Chapter 4:

$$\begin{gathered}\sigma_{ij,kk} + \frac{m'}{m'+1}\left(\sigma_{kk,ij} - \sigma_{kk,ll}\delta_{ij}\right) - \frac{\rho}{\mu}\ddot{\sigma}_{ij} + \\ + \frac{\rho(\lambda+\mu)}{\mu(3\lambda+2\mu)}\ddot{\sigma}_{kk}\delta_{ij} + \rho\frac{\partial}{\partial t}\left(j_{ij}+j_{ji}\right) - \rho\frac{\partial}{\partial t}j_{ll}\delta_{ij} = 2\mu\eta_{ij}.\end{gathered} \tag{2.3}$$

Let us determine the elastic stresses and electric fields accompanying the bending of the domain wall in the crystal of ferroelectric–ferroelastic in the dynamic case. As in Chapter 4, let us assume for determinacy that the direction of spontaneous shear is perpendicular to vector $\mathbf{P}_0$ and coincides with axis $y$. In this case, the non-zero component of the tensor of spontaneous distortion is

$$u^s_{12} = -\epsilon_0\left[1 - 2\Theta\left(x - U(z,y)\right)\right]. \tag{2.4}$$

The components of the tensor of the density of twinning dislocations and, together with them, the non-zero components of the incompatibility tensor $j_{ij}$, evidently coincide with the corresponding expressions (3.4) and (3.5) of Chapter 4 in the problem of static bending of the ferroelastic domain wall. On the basis of definition of the tensor of the density of the flow of twinning dislocations (expression (2.14) in Chapter

4) and equation (2.4) in this chapter, tensor $j_{ij}$ has the unique component differing from zero

$$j_{12} = -2\epsilon\delta(x)\cdot\frac{\partial U}{dt}. \tag{2.5}$$

Let every element of the domain boundary make small harmonic vibrations that propagate along the wall in the form of a wave

$$U = U_0\cdot\exp(i\mathbf{k}\rho - \omega t),\ \mathbf{k} = (k_y, k_z),\ \rho(y,z) \tag{2.6}$$

Then

$$\begin{aligned}\sigma_{12} &= \tilde{\sigma}_{12}(x)\cdot\exp(i\mathbf{k}\boldsymbol{\rho} - \omega t),\\ \sigma_{kk} \equiv \sigma &= \tilde{\sigma}(x)\cdot\exp(i\mathbf{k}\boldsymbol{\rho} - \omega t),\\ \varphi &= \tilde{\varphi}(x)\cdot\exp(i\mathbf{k}\boldsymbol{\rho} - \omega t).\end{aligned} \tag{2.7}$$

On the basis of the Beltrami equation, the expressions for the components of the tensor $\eta_{ij}$ and component $j_{12}$ (2.5), expressed by means of displacement of the wall $U$, and also by means of representations (2.6)–(2.7) the equation for component $\sigma_{12}$ is as follows

$$\sigma_{12}'' - k^2\sigma_{12} + \beta i k_y \sigma' - \frac{\rho}{\mu}\ddot{\sigma}_{12} = 2\mu\epsilon_0 k_z^2 U + 2\rho\epsilon_0\delta(x)\frac{\partial^2 U}{\partial t^2}. \tag{2.8}$$

The system of equations for $\sigma_{11}$, $\sigma_{22}$, $\sigma_{33}$ is

$$\begin{aligned}&\sigma_{11}'' - k^2\sigma_{11} + \beta k^2\sigma - \frac{\rho}{\mu}\ddot{\sigma}_{11} + \frac{\rho}{\mu}\frac{(\lambda+\mu)}{(3\lambda+2\mu)}\ddot{\sigma} = 0,\\ &\sigma_{22}'' - k^2\sigma_{22} + \beta(\sigma k_z^2 - \sigma'') - \frac{\rho}{\mu}\ddot{\sigma}_{22} + \frac{\rho}{\mu}\frac{(\lambda+\mu)}{(3\lambda+2\mu)}\ddot{\sigma} = 0,\\ &\sigma_{33}'' - k^2\sigma_{33} + \beta(\sigma k_y^2 - \sigma'') - \frac{\rho}{\mu}\ddot{\sigma}_{33} + \frac{\rho}{\mu}\frac{(\lambda+\mu)}{(3\lambda+2\mu)}\ddot{\sigma} =\\ &= -4\mu\epsilon_0 i k_y U\delta'(x).\end{aligned} \tag{2.9}$$

Summing them up, we obtain

$$(1-2\beta)\sigma'' - k^2(1-2\beta)\sigma + \frac{\rho}{(3\lambda+2\mu)}\ddot{\sigma} = -4\mu\epsilon_0 i k_y U\delta'(x). \tag{2.10}$$

Using the Fourier-expansion

$$\tilde{\sigma}(x) = \int_{-\infty}^{\infty}\sigma_{k_x}\exp(ik_x x)\frac{dk_x}{2\pi}, \tag{2.11}$$

we obtain

$$\sigma_{k_x} = \frac{4\mu\epsilon_0 k_y k_x U_0}{(2\beta-1)\left[k_x^2+k^2-\rho\omega^2/(3\lambda+2\mu)(2\beta-1)\right]}. \tag{2.12}$$

By means of the Fourier-expansion $\sigma_{12}$, on the basis of equation (2.8) and expression (2.12) we have

$$\sigma_{12}^{k_x} = -\frac{2\mu\epsilon_0 U_0\left(k_z^2-\rho\omega^2/\mu\right)}{\left[k_x^2+k^2-\rho\omega^2/\mu\right]} -$$

$$-\frac{4\mu\epsilon_0 k_y^2 k_x^2 \beta U_0}{(2\beta-1)\left[k_x^2+k^2-\rho\omega^2/(3\lambda+2\mu)(2\beta-1)\right]\left[k_x^2+k^2-\rho\omega^2/\mu\right]}. \tag{2.13}$$

Hence

$$\sigma_{12}\Big|_{x=0} = \frac{-\mu\epsilon_0 U\left(k_z^2-\rho\omega^2/\mu\right)}{\sqrt{k^2-\rho\omega^2/\mu}} -$$

$$-\frac{2\mu\epsilon_0 U\beta k_y^2}{(2\beta-1)\left[\sqrt{k^2-\rho\omega^2/\mu}+\sqrt{k^2-\rho\omega^2/(\lambda+2\mu)}\right]}. \tag{2.14}$$

Similarly, on the basis of the Poisson equation (2.2) for the tetragonal symmetry of tensor $\varepsilon_{ij}$ we determine the potential of the bound charges on the boundary

$$\varphi = \frac{-4\pi P_0 i k_z U}{\varepsilon_a\sqrt{\frac{\varepsilon_c}{\varepsilon_a}k_z^2+k_y^2}} \exp\left[-|x|\sqrt{\frac{\varepsilon_c}{\varepsilon_a}k_z^2+k_y^2}\right] \tag{2.15}$$

and

$$\frac{\partial\varphi}{\partial z}\Big|_{x=0} = \frac{4\pi P_0 k_z^2 U}{\varepsilon_a\sqrt{\frac{\varepsilon_c}{\varepsilon_a}k_z^2+k_y^2}}. \tag{2.16}$$

Substituting (2.14) and (2.16) into (2.1) after cancelling the common factor $U$ we have

$$m\omega^2 = \frac{2\mu\epsilon_0^2\left(k_z^2 - \rho\omega^2/\mu\right)}{\sqrt{k^2 - \rho\omega^2/\mu}} +$$
$$+\frac{4\mu\epsilon_0^2\beta k_y^2}{(2\beta-1)\left[\sqrt{k^2 - \rho\omega^2/\mu} + \sqrt{k^2 - \rho\omega^2/(\lambda+2\mu)}\right]} +$$
$$+\frac{8\pi P_0^2 k_z^2}{\varepsilon_a\sqrt{\frac{\varepsilon_c}{\varepsilon_a}k_z^2 + k_y^2}}. \tag{2.17}$$

Taking into account that $c_t^2 = \mu/\rho$, $c_l^2 = (\lambda + 2\mu)/\rho$ and introducing ratio $\omega = \upsilon \cdot k$, where $\upsilon$ is the propagation velocity of the wave of bending displacements of the domain wall, for small enough $k$, when it is possible to ignore the local mass of the domain wall, and taking into account the expression for $\beta$, we can rewrite equation (2.17) in the form

$$\frac{\left(\cos^2\varphi - \upsilon^2/c_t^2\right)}{\sqrt{1-\upsilon^2/c_t^2}} + 4\left(1-\frac{c_t^2}{c_l^2}\right)\cdot\frac{\sin^2\varphi}{\left[\sqrt{1-\upsilon^2/c_t^2} + \sqrt{1-\upsilon^2/c_t^2}\right]} +$$
$$+\frac{\tilde{\gamma}\cdot\cos^2\varphi}{\sqrt{\frac{\varepsilon_c}{\varepsilon_a}\cos^2\varphi + \sin^2\varphi}} = 0, \tag{2.18}$$

$$\tilde{\gamma} = 4\pi P_0^2/\varepsilon_a\mu\epsilon_0^2,$$

where angle $\varphi$ is counted from the polar direction or, after evident transformations, as in [197, 198]

$$4\sqrt{1-\frac{\upsilon^2}{c_t^2}}\cdot\sqrt{1-\frac{\upsilon^2}{c_l^2}} - \left(2-\frac{\upsilon^2}{c_t^2}\right)^2 + \frac{\upsilon^2}{c_t^2}\left(1-\frac{\upsilon^2}{c_t^2}\right)\cdot\mathrm{ctg}^2\varphi +$$
$$+\tilde{\gamma}\frac{\upsilon^2}{c_t^2}\sqrt{1-\frac{\upsilon^2}{c_t^2}}\cdot\cos^2\varphi = 0. \tag{2.19}$$

Analysis of the obtained equation shows the following. In the pure ferroelastic ($P_0 = 0$) the propagation of the wave of bending displacements of the domain wall is described by equation (2.19) without the last term ($\gamma = 0$). It is clearly seen that for the given specific direction, the velocity of propagation of the wave is constant and independent of $k$. This velocity depends on the direction of the propagation of the wave, i.e. $\upsilon = \upsilon(\varphi)$. The orientation dependence of $\upsilon$ is such that for the direction of spontaneous shear ($\varphi = \pi/2$) it coincides with the velocity of the Rayleigh wave, which is determined by the equation [99, 199]

$$\left(2-\frac{\upsilon^2}{c_t^2}\right)^2 = 4\sqrt{1-\frac{\upsilon^2}{c_t^2}}\sqrt{1-\frac{\upsilon^2}{c_l^2}}. \tag{2.20}$$

In the case of deviation from this direction, the wave is gradually transformed and completely changes to a volume shear wave for direction $\varphi = 0$ normal to the direction of spontaneous shear (Fig. 7.2).

In terms of twinning dislocations, the dynamics of corresponding displacements is determined by the interaction of an ensemble of moving dislocations. At $\varphi = \pi/2$ these are purely edge dislocations. At $\varphi < \pi/2$ screw dislocations add up to them. With the appearance of these dislocations, the velocity of the surface wave on the domain wall increases and its localization decreases respectively, tending to infinity for the volume shear wave (Fig. 7.2) at $\varphi = 0$.

In a defect-free ferroelectric–ferroelastic the limiting values for the velocity of the boundary surface waves at $\varphi = 0$ and $\varphi = \pi/2$ remain the same. The value of $\upsilon$ for intermediate values of $\varphi$ in comparison with the pure ferroelastic is always higher. Evidently, this is caused by the appearance here of the additional rigidity of the boundary in relation to its bending displacements, and connected to polarization.

As in the case of the pure ferroelectric, the direct proportionality

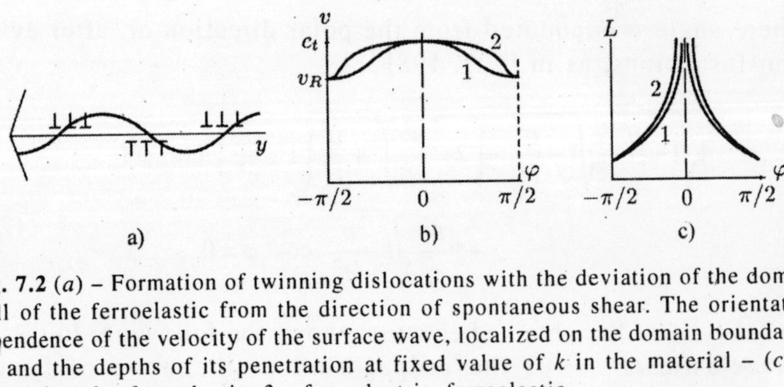


**Fig. 7.2** (*a*) – Formation of twinning dislocations with the deviation of the domain wall of the ferroelastic from the direction of spontaneous shear. The orientation dependence of the velocity of the surface wave, localized on the domain boundary – (*b*) and the depths of its penetration at fixed value of $k$ in the material – (*c*) in defect-free *1* – ferroelastic, *2* – ferroelectric–ferroelastic.

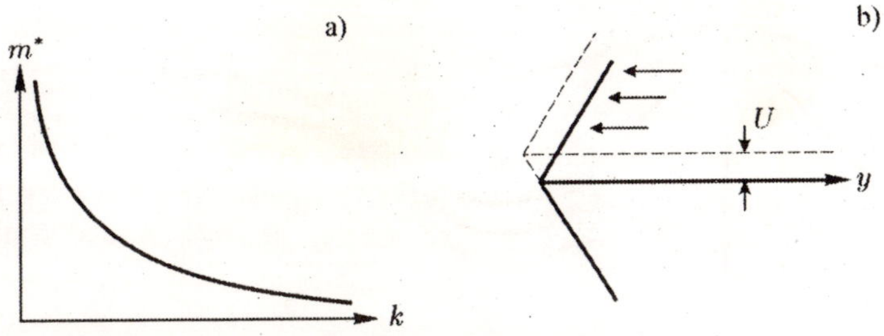


**Fig. 7.3** Dependence of the effective mass of a domain wall on the wave vector in pure ferroelectrics (excluding the case with $k = 0$), ferroelastics and ferroelectrics–ferroelastics (*a*). Formation of an infinite mass (the involvement into movement of all upper half-space) at translational ($k = 0$) displacement of the domain boundary in a ferroelastic (*b*).

$\omega \sim k$ in defect-free ferroelastics and ferroelectrics–ferroelastics is associated with the involvement in the motion of the inertia medium surrounding the boundary (Fig. 7.3), with the thickness of the layer of $\sim 1/k$. Together with the proportionality to $k$ of the coefficient of the quasi-elastic force acting on the boundary, which is associated with the displacement of the boundary (the right-hand part of (2.17) at $\omega = 0$), all this results in the linear dependence of $\omega$ on $k$.

It should be mentioned that in contrast to the pure ferroelectric with the 180-degree domain structure, where the displacement of the domain boundary as a whole in the infinite material does not result in formation of electric fields and, consequently, in the piezoelectric deformation of the material, i.e. $m^*(k=0)=0$ (point $k=0$ is the unique point in this case), and there is no direct elastic interaction, in the case of the ferroelastic, as shown in Fig. 7.3, the translational motion of the domain wall results in infinite increase of its effective mass, i.e. $m^*(k=0)$ is equal to infinity in this case.

As in the ferroelectric–ferroelastic, the 90° domain walls in the ferroelectrics of the perovskite type in particular, separate the domains, which differ not only in spontaneous polarization but also in spontaneous deformation. This is caused by electrostrictive interactions, which in the laboratory system of coordinates rotated by 45° around the stationary axis $y$ and by135° with respect to the crystallographic systems, separated by the domain wall (Fig. 7.4), lead to the formation of spontaneous deformation [5]

$$\epsilon_0 = u_{13}^s = \frac{q_{33} - q_{32}}{c_{33} - c_{32}} [P_{03}] \; P_{01} = \frac{q_{33} - q_{32}}{c_{33} - c_{32}} P_0^2, \qquad (2.21)$$

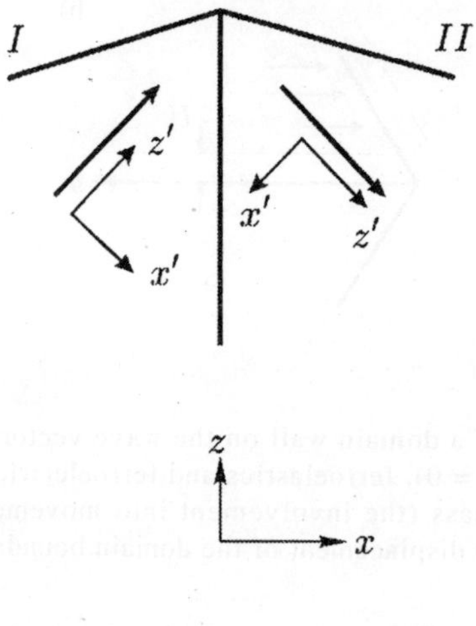


**Fig. 7.4** Mutual orientation disposition of spontaneous polarisation, the crystallographic systems of coordinates of the adjacent 90°-domains and the laboratory system of coordinates.

where the following notation was used

$$[P_{03}] = P_{03}^{II} - P_{03}^{I} = -P_0/\sqrt{2}, \tag{2.22}$$

the indices I and II denote the adjacent domains and $q_{\alpha\beta}$, $c_{\alpha\beta}$ are the tensors of electrostriction constants and elastic moduli respectively, written in the abbreviated index notations.

As indicated by the numeration of the indices of the ferroactive component $u_{13}^s$ in (2.21), in contrast to the case of the ferroelastic–ferroelectric, where the direction of spontaneous shear is perpendicular to the vector $\mathbf{P}_0$, in this case these directions are parallel. Taking this into account and proceeding to writing the tensor of dielectric constants in the laboratory system of co-ordinates instead of the crystallographic system, as in the case of the ferroelectric–ferroelastic the same linear dependence of $\omega$ on $k$ is obtained here in the law of dispersion of bending vibrations of the 90° domain walls in the ferroelectric where the velocity of their propagation is determined by the equation similar to (2.18) [200]:

$$\frac{\left(\cos^2\varphi - v^2/c_t^2\right)}{\sqrt{1-v^2/c_t^2}} + 4\left(1-\frac{c_t^2}{c_l^2}\right)\cdot\frac{\sin^2\varphi}{\left[\sqrt{1-v^2/c_t^2}+\sqrt{1-v^2/c_l^2}\right]} + \\ + \frac{\tilde{\gamma}\cdot\sin^2\varphi}{\sqrt{\cos^2\varphi+\delta\cdot\sin^2\varphi}} = 0. \tag{2.23}$$

Hence $\tilde{\gamma} = 4\pi P_0^2/\mu\epsilon_0^2\sqrt{\varepsilon_c\varepsilon_a}$, $\delta = (1+\varepsilon_a/\varepsilon_c)/2$.

As it follows from the analysis of (2.23), the presence here in the last term on the left in contrast to (2.18) of $\sin^2\varphi$, and not $\cos^2\varphi$,

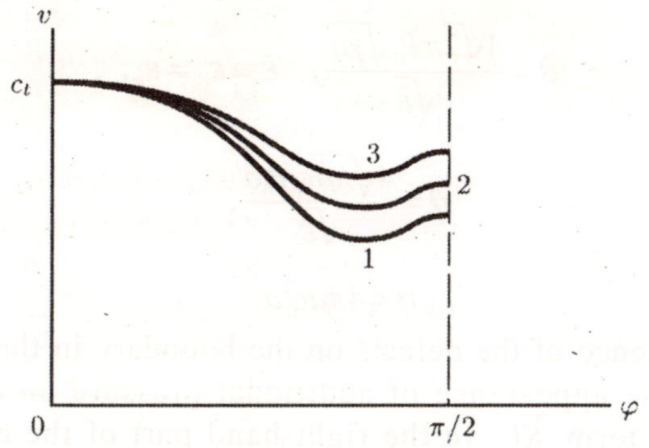


**Fig. 7.5** Orientation dependence of the velocity of the surface wave on the 90°-domain wall in a ferroelectric: $\tilde{\gamma}_1 < \tilde{\gamma}_2 < \tilde{\gamma}_3$.

related to the alteration of the mutual orientation of the spontaneous polarization vector and the direction of spontaneous shear, results in the following differences of the waves on the 90° domain wall in the ferroelectric from the previously examined cases. Firstly, the velocity of these waves is not always a monotonic function of angle $\varphi$ (Fig. 7.5). Secondly, due to the presence of additional rigidity of the domain wall of the electrostatic origin, which does not disappear along the direction of spontaneous shear, the purely Rayleigh wave is not realized here for any direction of propagation. For the same reason, as in the ferroelectric–ferroelastic, the velocity of the surface wave localized on the 90° domain wall, is always higher than in the case of the pure ferroelastic for the same values of elastic constants.

## 7.3. BENDING VIBRATIONS OF DOMAIN WALLS OF REAL FERROELECTRICS AND FERROELASTICS

The presence of crystalline lattice defects in a crystal leads, as shown later, to qualitative changes in the spectrum of vibration of the domain boundaries, changing not only the velocity of propagation of the wave of bending displacement of the boundaries but also the dispersion law of their vibrations itself. To determine this law when describing the influence of defects in the long-wave approximation, it is possible to use, as it was done in Chapter 5, assumptions about the quasi-elastic force, acting on the boundary from the direction of the defect. For isolated defects in the crystals of the pure ferroelectric with 180° domains, ferroelectric–ferroelastic and pure ferroelastic the coefficients of quasi-elasticity of the corresponding forces are equal to

$$\vartheta = \frac{4\sqrt{2}\pi P_0 \sqrt{\gamma a}}{\sqrt{\tilde{\varepsilon}}}, \quad \tilde{\varepsilon} \equiv \varepsilon_c = \varepsilon_a, \tag{3.1}$$

$$\vartheta = \frac{4\sqrt{\pi\mu\epsilon_0} P_0 a}{\sqrt{\tilde{\varepsilon}}}, \tag{3.2}$$

$$\vartheta = 4\pi\mu\epsilon_0^2 a. \tag{3.3}$$

The influence of the defects on the boundary in this approximation results in the appearance of additional pressure on the boundary in the form of term $\overline{K}U$ in the right-hand part of the equations of the boundary motion (1.6) and (2.1), where $\overline{K} = \vartheta / l^2$ and $l$ is the average distance between the defects pinning the boundary and, consequently, it leads to the direct addition of the coefficient $\overline{K}$ to the right-hand part of equations (1.17) and (2.18) describing the laws of dispersion of the bending vibrations of the domain boundaries in pure ferroelectrics and ferroelectrics–ferroelastics.

The result of the influence of the defects on the dependence $\omega(\mathbf{k})$ under consideration is already clearly visible from the example of pure ferroelectrics with the 180° domain structure. In the presence of defects, the equation describing the law of dispersion of the bending vibrations of the domain boundaries has the form

$$\left( m + \frac{5\pi^2 P_0^2 \beta_1^2 k_z^6}{\rho \cdot \varepsilon^2 c_l^4 \left(k_x^2 + k_y^2\right)^{7/2}} \right) \omega^2 = \frac{4\pi P_0^2 k_z^2}{\varepsilon_a \sqrt{\frac{\varepsilon_c}{\varepsilon_a} k_x^2 + k_y^2}} + \overline{K}. \tag{3.4}$$

As it can be seen from equation (3.4) and Fig. 7.6, the presence of defects results in a change of the effective coefficient of quasi-

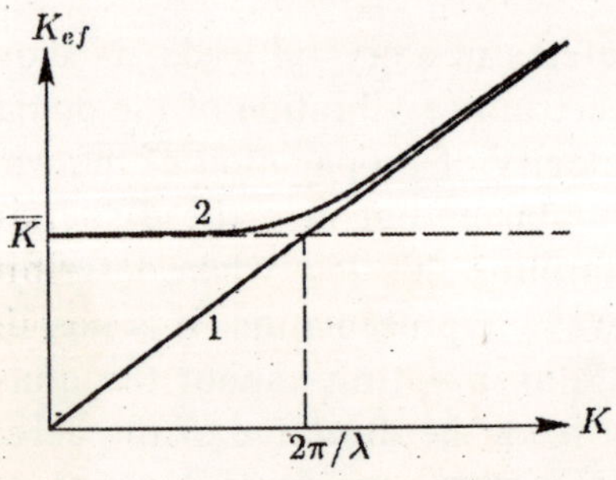


**Fig. 7.6.** Dependence of the effective coefficient of the quasielastic force, acting on the boundary, on the wave vector in a defect-free (a) and defective (2) materials.

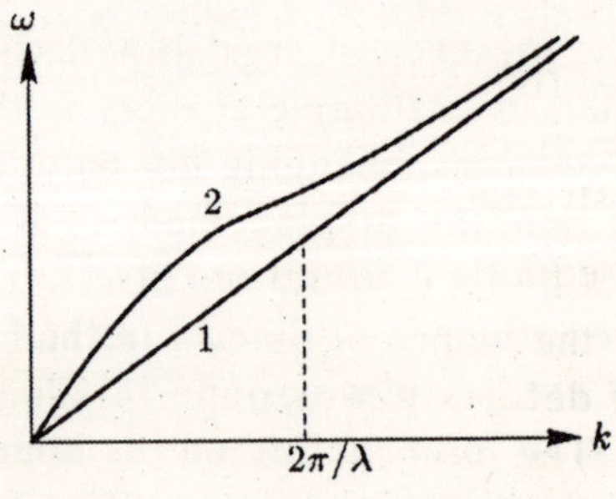


**Fig. 7.7.** The law of dispersion of bending vibrations of domain boundaries in a defective material (2) in comparison with a defect-free material (1).

elasticity, which is now determined by two terms. For the polar direction in the range of low values of $k$, where the rigidity of the boundary in the defect-free material decreases in proportion to $k$, the behaviour of $K_{ef}$ is determined by the term $\overline{K}$. In the range of high $k$, the situation is opposite.

As the result, the linear dependence $\omega \sim k$, which in the defect-free material was observed both in the range of low and relatively high $k$, in the crystal with defects will be implemented only at high $k$. In the long-wave limit the dependence $\omega(\mathbf{k})$ is transformed to the route dependence $\omega \sim \sqrt{k}$. For polar direction in particular it takes place at $k_z = \overline{K}\sqrt{\varepsilon_a \varepsilon_c}/4\pi P_0^2$. Taking into account the ratio of the coefficient $\overline{K}$ with the individual coefficient $\vartheta$ and the direct expression for the latter (3.1) we can find the value of the critical wavelength $\lambda$, at which the change of the vibration modes takes place (see also Fig. 7.7) [194, 195]:

$$\lambda = \frac{\sqrt{2}\pi P_0 l^2}{(\varepsilon_c \varepsilon_a)^{1/4}\sqrt{\gamma a}}. \tag{3.5}$$

For conventional $P_0 \sim 10^4$, $\sqrt{\varepsilon_c \varepsilon_a} \sim 10^3$, $a \sim 10^{-7}$cm, $\gamma \sim 0.1$ erg·cm$^{2}$, $l \sim 10^{-6}$cm (the concentration of defects $\sim 10^{18}$cm$^{-3}$) it takes place at $\lambda \sim 10^{-5}$cm.

The similar situation with the change of the vibration mode will be observed for all other directions of vector $\mathbf{k}$, with the exception of the direction perpendicular to the polar axis. At that the critical value of $\lambda$ will continuously decreases with the deviation of the vector $\mathbf{k}$ from the polar direction.

The direction perpendicular to vector $\mathbf{P}_0$ is special. For this direction there is an eigenfrequency of vibrations of the domain boundaries, which is almost independent of $k$ up to $k = \sqrt{\overline{K}/\gamma}$ (at conventional $l \sim 10^{-6}$cm this value of $k$ is approximately equal to $3 \times 10^6$cm$^{-1}$, i.e. it is almost on the upper limit of the permissible values of $k$, restricted by the approximation of the structureless boundary) and is equal to $\omega = \sqrt{\overline{K}/m}$. As it can be seen from the determination of $\overline{K}$ the value of the mentioned frequency depends on the concentration of defects interacting with the boundary.

The addition of the term associated with the defects in the case of the ferroelastic–ferroelectric shows that in comparison with the ideal material, the dependence of $\omega$ on $k$ becomes non-linear. Nevertheless, it is convenient to carry out its analysis if it is assumed that as before $\omega = \upsilon k$ with the velocity of propagation already dependent on $k$. After all transformations this results in the following equation that determines the dependence of $\upsilon(\mathbf{k})$ for $\varepsilon_c = \varepsilon_a$ (and hence as the result $\omega(\mathbf{k})$ in the given case [194, 195]:

$$4\sqrt{1-\frac{\upsilon^2}{c_t^2}}\sqrt{1-\frac{\upsilon^2}{c_l^2}}-\left(2-\frac{\upsilon^2}{c_t^2}\right)^2+\frac{\upsilon^2}{c_t^2}\left(1-\frac{\upsilon^2}{c_t^2}\right)\mathrm{ctg}^2\,\varphi+$$
$$+\frac{K}{2\mu\epsilon_0^2}\frac{1}{k}\frac{\upsilon^2}{c_t^2}\sqrt{1-\frac{\upsilon^2}{c_t^2}\frac{1}{\sin^2\varphi}}+\tilde{\gamma}\frac{\upsilon^2}{c_t^2}\sqrt{1-\frac{\upsilon^2}{c_t^2}\mathrm{ctg}^2\,\varphi}=0, \tag{3.6}$$

where, as previously, angle $\varphi$ is counted from the polar direction.

Analysis of (3.6) shows that, like in the pure ferroelectric with defects, in the real (defective) ferroelectric–ferroelastic and also in the pure ferroelastic, for all directions, with the exception of the direction of spontaneous shear, the frequency of bending vibrations of the domain boundaries increases in the range of low values of $k$. At that, the velocity of the corresponding wave tends to $c_t$ at $k \to 0$. Taking equation (3.2) into account the conventional linear dependence $\omega(\mathbf{k})$ for the direction of spontaneous shear in particular in ferroelectric–ferroelastic is achieved at

$$\lambda = \frac{\sqrt{\mu\tilde{\varepsilon}}\,\epsilon_0 l^2}{P_0 a}. \tag{3.7}$$

In the direction perpendicular to spontaneous shear and in ferroelectric–ferroelastic where the direction of spontaneous shear is $\mathbf{s}\perp\mathbf{P}_0$, and in the pure ferroelastic, both in the defective material and in the ideal crystal, the volume shear wave propagates at all values of $k$.

## 7.4. TRANSLATIONAL VIBRATIONS OF THE DOMAIN STRUCTURE IN FERROELECTRICS AND FERROELASTICS

The factors causing the formation of bending dynamics of domain boundaries also lead to the formation of translational vibrations of the domain structure, i.e. such vibrations at which the domain boundaries are displaced as a whole unit in the direction normal to the polar axis. The mentioned displacements of the domain boundaries also lead to changes in the distribution of charges of spontaneous polarization, but now on the surface of the material, in the area of contact of the material with the surface non-ferroelectric layer or with a block of the domain structure of a different orientation, thus causing an increase of the electrostatic energy of the polydomain ferroelectric and, therefore, the appearance of the restoring forces acting on the displaced domain boundaries [201,223]. For the 90° domain boundaries in ferroelectrics and also in the case of ferroelectrics–ferroelastics in the presence of mechanical contact of the domain structure with other part of the material, not experiencing such displacements as the domain structure under consideration, the displacements of the domain boundaries cause the increase of the mechanical energy of the system and, therefore, the appearance of the restoring forces of the corresponding nature. As in the case of bending vibrations of the domain boundaries, the other fundamental factor for the formation of translational vibrations of the domain structure are the inertial properties of the domain boundaries and of the ferroactive material itself.

The principle considerations of such vibrations will be carried out using the example of a pure ferroelectric with the 180° domain structure. Let us formulate a set of equations determining the type of the domain boundaries motion in this case. Let us determine the system of coordinates, as previously, in such a manner that axis $z$ coincides with the polar direction, and axis $x$ with the direction of motion of the domain boundaries. Let us assume that in the process of translational displacements of the domain walls the latter remain flat (Fig. 7.8) and the value of the displacements $U_n \ll d$ ($n$ is the number of the walls, $d$ is the average width of the domain). Consequently, in the harmonic limit, the electrostatic energy of the ferroelectric material

$$\Phi = \frac{L_y}{2} \int P(x,z) \frac{d\varphi}{dz} \, dxdz, \tag{4.1}$$

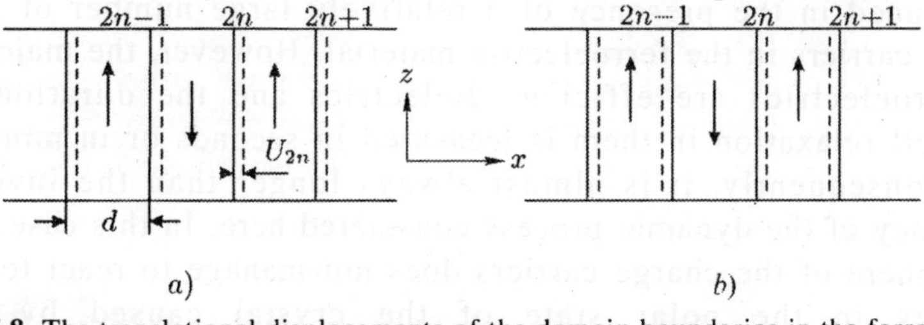


**Fig. 7.8.** The translational displacements of the domain boundaries in the ferroelectric with the 180° domain structure in the quasi-acoustic (a) and quasi-optical branches of vibrations of the domain structure (b).

can be conveniently presented in the form of the expansion

$$\Phi = \Phi_0 + \frac{1}{2}\sum_{n,n'} \frac{\partial^2 \Phi}{\partial U_n \partial U_{n'}}\bigg|_0 U_n U_{n'} = \Phi_0 + \frac{1}{2}\sum_{n,n'} \tilde{K}(n,n')\, U_n U_{n'}. \tag{4.2}$$

Here $\Phi_0$ is the energy of the depolarizing field for the nondisplaced domain boundaries, i.e. for their periodic distribution, which corresponds to the equilibrium domain structure, $L_y$ is the size of the crystal in the corresponding direction.

On the basis of equation (4.2), the equation of motion of the $n$-th domain wall

$$\tilde{m}\ddot{U}_n = -\frac{\partial \Phi}{\partial U_n} \tag{4.3}$$

takes the following form

$$\tilde{m}\ddot{U}_n + \tilde{K}(n,n)U_n + \sum_{n \neq n'} \tilde{K}(n,n')U_{n'} = 0. \tag{4.4}$$

Here $\tilde{m} = mL_y L_z$, $m$ is the density of the local effective mass of the domain wall.

A special feature of this problem, as well as of the problem of bending vibrations of the domain boundaries in ferroelectrics, is the involvement in the motion due to displacements of the domain boundaries as a result of the piezoelectric effect of the entire bulk of the material, where the field of spontaneous polarization charges that changes in the process of the domain boundaries motion is localized. Therefore, in calculation of the coefficients of quasi-elastic forces $\tilde{K}(n,n')$, acting on the domain walls and displaced from the equilibrium positions in (4.4), the electrostatic equation must also be supplemented here by the equation of the motion of the elastic medium and the appropriate material equations in the crystal with the piezoelectric effect.

It should be noted that the role of long-range electric fields in controlling the displacement of the domain boundaries in general can be reduced in the presence of a relatively large number of free charge carriers in the ferroelectric material. However, the majority of ferroelectrics are efficient dielectrics and the duration of Maxwell relaxation in them is measured in seconds or in minutes and, consequently, it is almost always longer than the inverse frequency of the dynamic process considered here. In this case, the atmosphere of the charge carriers does not manage to react to the changes in the polar state of the crystal caused by the displacements of the domain boundaries. And since for the identical displacements of the domain walls the changes in the total density

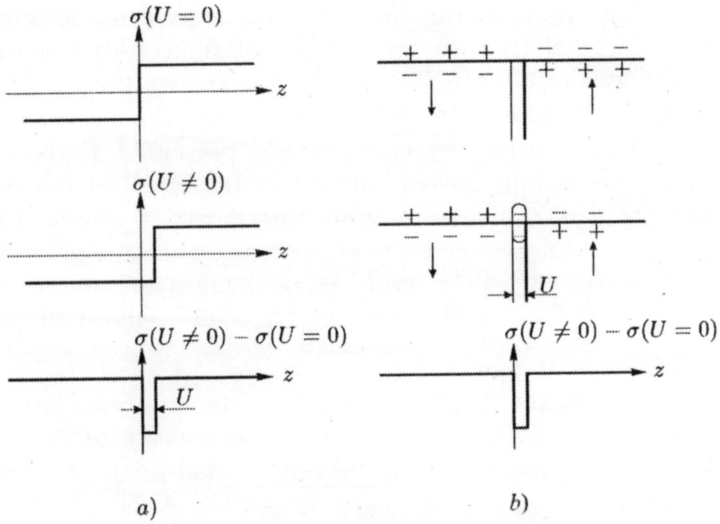


**Fig. 7.9.** Equivalence of the resultant charge for equal displacement of the domain boundaries for the following cases: initially non-compensated (a) and completely compensated (b) charges of spontaneous polarization.

of the charge, which actually control their displacements are practically identical (Fig. 7.9), regardless of whether the long-range electric field in the initial state has or has not been compensated, the influence of the free carriers in this case can be disregarded in specific calculations.

As it was mentioned previously, the displacements of the domain walls in this problem are assumed to have no bending and that is why the surface of the ferroelectric material is the only place for location of spontaneous polarization charges. We will consider the infinite ferroelectric plate with the thickness of $L_z$ along the polar axis. Let us choose the origin of the coordinates in the middle of its thickness and coinciding with one of the walls along the direction of the infinite length of the ferroelectric, axis $x$. Let us ignore the coordinate dependence of the value of spontaneous polarization in the vicinity of the surface of the ferroelectric plate along the direction of the polar axis, assuming that it has a constant (independent of $z$) value equal to $\pm P_0$ within the limits of the domain of a specific sign and drops to zero stepwise on the surfaces of the ferroelectric plate when the coordinates are equal to $\pm L_z/2$. It is also assumed that the domain boundary is structureless, i.e. has zero thickness. This means that the bound charge on the surface of the ferroelectric will be distributed in steps with constant values

$\pm P_0$ within the limits of the domain of the appropriate orientation, i.e. in the interval from the coordinate of some $n$-th domain wall $x_n = nd+U_n$ to the coordinate of the adjacent wall $x_{n\pm1}$.

The above distribution of the charge on the surface of the ferroelectric can further be conveniently presented as the sum of periodic distribution, corresponding to the equilibrium domain structure [5]

$$\sigma_0(x)=\begin{cases}-P_0, & (2n-1)d<x<2nd\\ P_0, & 2nd<x<(2n+1)d\end{cases} \tag{4.5}$$

at $z = L_z/2$ and antisymmetric to it at $z=-L_z/2$ and its variation $\sigma_1(x)$, linked with the displacement of the domain boundaries from the equilibrium positions. For small displacements of the domain boundaries $U_n \ll d$ the latter can be represented in the form [203]

$$\sigma_1(x)=\sum_n \gamma_n \delta(x-d\cdot n), \tag{4.6}$$

where the value $\gamma_n=2P_0U_n$ at $z=L_z/2$ and is equal to $-2P_0U_n$ at $z=L_z/2$.

The equation for the electric field of the given distribution of the charges (4.5), (4.6) after substitution of ratios (1.2) and (1.3) into the electrostatic equation (1.2) is a heterogeneous equation, which can be solved using the method of the Green function. The set of equations for determination of its Fourier image will be obtained from these equations after replacing here the real coordinate distribution of the changes of spontaneous polarization $\partial P_{0i}/\partial x_i$ by $\delta(x,z)$–like source. Taking into account the expression for the strain tensor, the relation of the electrostatic field $E_i$ with the potential $\varphi$ and the axial symmetry of the problem, the set has the form

$$\begin{cases}-\varepsilon_{ij}\dfrac{\partial^2\varphi}{\partial x_i \partial x_j}-4\pi\beta_{ijk}\dfrac{\partial^2 u_j}{\partial x_k \partial x_i}=4\pi\gamma(t)\delta(x)\delta(z),\\[2ex] \rho\ddot{u}_i=c_{ijkl}\dfrac{\partial^2 u_k}{\partial x_l \partial x_j}-\beta_{kij}\dfrac{\partial^2\varphi}{\partial x_k \partial x_j},\end{cases} \tag{4.7.}$$

where in accordance with the conditions of the problem $\gamma=\gamma_0\cdot\exp(i\omega t)$.

The further determination of the relation between potential $\varphi$ and displacement $u_i$ from (4.7) is similar to the case of bending vibrations of the domain boundaries in a pure ferroelectric. Representing $\varphi(x,z)$ and $u_i(x,z)$ in the form of Fourier expansion:

$$\varphi(x,z)=\int\varphi_{\mathbf{k}}(t)e^{-ik_x x}e^{-ik_z z}\frac{dk_x dk_z}{(2\pi)^2},$$

$$u_i(x,z)=\int u_{i\mathbf{k}}(t)e^{-ik_x x}e^{-ik_z z}\frac{dk_x dk_z}{(2\pi)^2}, \tag{4.8}$$

where $\varphi_{\mathbf{k}}(t)$ and $u_{i\mathbf{k}}(t)$ are proportional to $\exp(i\omega t)$, for the material isotropic in the elastic respect as in the considerations of section 7.1. for the normalized Fourier coefficient $\tilde{\varphi}_{\mathbf{k}}=\varphi_{\mathbf{k}}/\gamma$, which represents the Fourier image of the Green function of the equation for the electric field taking the piezoelectric effect into account, we obtain

$$\tilde{\varphi}_{\mathbf{k}}(t)=4\pi\left\{\varepsilon_{ij}k_i k_j+\frac{4\pi\beta_{ijk}k_k k_i}{\rho\left(c_t^2\tilde{k}^2-\omega^2\right)}\times\right.$$

$$\left.\times\left[\beta_{pmj}k_p k_j-\frac{\left(c_l^2-c_t^2\right)}{\left(c_l^2\tilde{k}^2-\omega^2\right)}\cdot\beta_{plj}k_p k_j k_l k_m\right]\right\}^{-1}, \tag{4.9}$$

where $\tilde{k}^2=k_x^2+k_z^2$.

The interaction of the equilibrium distributed charges (4.5) with each other determines in (4.2) only the constant term $\Phi_0$ which is not included in the equation of domain boundaries motion (4.4). The next term in (4.2) is determined by the interaction of charges (4.6). Therefore, it is natural here to restrict our considerations by the calculation of the fields and interaction of only these charges in particular. The volume density of the charge, corresponding to (4.6), distributed on both surfaces of the ferroelectric plate is

$$\rho(x,z)=\sum_n \gamma_n\delta(x-dn)\left[\delta\left(z-\frac{L_z}{2}\right)-\delta\left(z+\frac{L_z}{2}\right)\right]. \tag{4.10}$$

Then, in accordance with the properties of the Green function, the potential of these charges is

$$\varphi_1(x,z)=\int\rho_{\mathbf{k}}\tilde{\varphi}_{\mathbf{k}}\cdot e^{-ik\rho}\cdot\frac{dk}{(2\pi)^2}=\sum_{n'}\gamma n'\int\tilde{\varphi}_{\mathbf{k}}e^{-ik_x(x-dn')}\times$$

$$\times\left[\exp\left(-ik_z\left(z-\frac{L_z}{2}\right)\right)-\exp\left(-ik_z\left(z+\frac{L_z}{2}\right)\right)\right]\times\frac{dk_x\cdot dk_z}{(2\pi)^2}. \tag{4.11}$$

The energy of their interaction

$$\Phi-\Phi_0=L_y\int\sigma_1\left(z=\frac{L_z}{2},x\right)\cdot\varphi_1\left(z=\frac{L_z}{2},x\right)dx=$$
$$=L_y\sum_n\sum_{n'}\gamma_n\gamma_{n'}\tilde{\varphi}_{\mathbf{k}}\cdot\exp\left(-ik_xd\left(n-n'\right)\right)\times$$
$$\times\left[1-\exp(-ik_zL_z)\right]\frac{dk_xdk_z}{\left(2\pi\right)^2}. \tag{4.12}$$

Taking into account that $\gamma=2P_0U_n$, and $\gamma_{n'}=2P_0U_{n'}(-1)^{n-n'}$, where the multiplier $(-1)^{n-n'}$ takes into account that at the same signs of $U_n$ and $U_{n'}$ the charge, appearing on the surface of the crystal, for the adjacent domain walls has the opposite sign, the last equation may be written in the form

$$\Phi-\Phi_0=\frac{1}{2}\sum_{n,n'}\tilde{K}\left(n,n'\right)U_nU_{n'}, \tag{4.13}$$

where

$$\tilde{K}\left(n,n'\right)\equiv\tilde{K}\left(n-n'\right)=\left(-1\right)^{n-n'}8P_0^2L_y\int\tilde{\varphi}_{\mathbf{k}}\exp(-ik_xd\left(n-n'\right))\times$$
$$\times\left[1-\exp(-ik_zL_z)\right]\frac{dk_xdk_z}{\left(2\pi\right)^2}. \tag{4.14}$$

The specific expression for these coefficients can be calculated analytically in the absence of the piezoelectric effect, where for the tetragonal symmetry of the tensor of dielectric permittivity

$$\tilde{\varphi}_{\mathbf{k}}\left(\beta=0\right)=\frac{4\pi}{\left(\varepsilon_ck_z^2+\varepsilon_ak_x^2\right)}. \tag{4.15}$$

In this case, as with the ferromagnetic with the 180° domain structure [204, 205],

$$\tilde{K}\left(n,n'\right)_{n\neq n'}=\tilde{K}\left(n-n'\right)=$$
$$=\frac{8P_0^2L_y}{\sqrt{\varepsilon_a\varepsilon_c}}\left(-1\right)^{n-n'}\ln\left[1+\left(\sqrt{\frac{\varepsilon_a}{\varepsilon_c}}\frac{L_z}{d}\frac{1}{\left(n-n'\right)}\right)^2\right]. \tag{4.16}$$

Equation (4.16) cannot be used for the case of $n'=n$ because the method of calculation of the coefficients (4.16) does not foresee the separation of the effect of the self-influence of the charges formed in the region of the displaced domain wall which naturally leads to an infinite increase of expression (4.16) at $n=n'$.

To calculate any of the coefficients $\tilde{K}(n,n')$ in sum (4.13) in

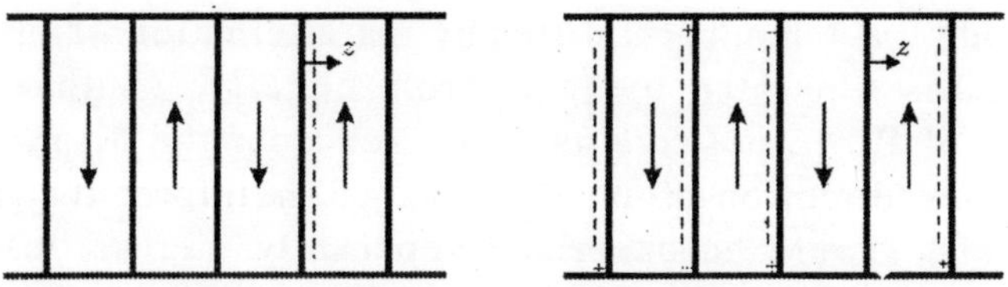


**Fig. 7.10.** The change of the electric state of the polydomain ferroelectric with the 180° domain structure when only one wall is displacement (a) and when all other walls are displaced the same distance (b).

principle it is necessary to displace from the initial position only the walls with the numbers $n$ and $n'$, leaving the others motionless so that sum (4.13) retains only one term. After that it is necessary to calculate the electrostatic energy of such a system $\Phi=\Phi_0+\frac{1}{2}\tilde{K}(n,n')U_nU_{n'}$ and separate in it the coefficient at $U_nU_{n'}$. Evidently, when calculating the coefficient $\widetilde{K}(n,n')$, it is necessary to displace only one domain wall (Fig.7.10 a) and calculate the change of the electrostatic energy of the system in this case.

As shown in Fig 7.10 b, it is like displacing all the other walls the same distance in the opposite direction, leaving the reference wall stationary. In this case, the origin of the coordinates does not coincide with any of the displaced walls and, consequently, for all of them it is possible to use coefficients (4.14) in calculation of the electrostatic energy. By summing up the terms $\widetilde{K}(n,n')\,U_nU_{n'}$ with $U_n=U_{n'}$ for all numbers $n$ and $n'$ starting with $|n'-n|=1$ to infinity for the situation in Fig.7.10 b and equating this energy change to the single term $\widetilde{K}(n,n')U_n^2$, we determine the coefficient

$$\tilde{K}(n,n)\equiv\tilde{K}=\frac{16P_0^2L_y}{\sqrt{\varepsilon_c\varepsilon_a}}\left\{\sum_{n=1}^{\infty}\ln\left[1+\frac{\varepsilon_a}{\varepsilon_c}\frac{L_z^2}{d^2}\frac{1}{(2n-1)^2}\right]-\right.$$
$$\left.-\sum_{n=1}^{\infty}\ln\left[1+\frac{\varepsilon_a}{\varepsilon_c}\frac{L_z^2}{d^2}\frac{1}{(2n)^2}\right]\right\}=$$
$$=\frac{16P_0^2L_y}{\sqrt{\varepsilon_c\varepsilon_a}}\ln\left(\frac{\pi}{2}\sqrt{\frac{\varepsilon_a}{\varepsilon_c}}\frac{L_z}{d}\right)\cdot\mathrm{cth}\left(\frac{\pi}{2}\sqrt{\frac{\varepsilon_a}{\varepsilon_c}}\frac{L_z}{d}\right). \qquad (4.17)$$

Let us examine the spectrum of vibrations of the domain structure in this case ($\beta=0$). Let the wall coinciding with the origin of the coordinates have the number $n$=0. As already mentioned, the adjacent walls (in the notations used, these are the walls with even

$2n$ and odd $2n-1$ numbers) differ by the alternation of the signs of the domains separated by them from negative (with a negative direction of $\mathbf{P}_0$) to positive and vice versa for the displacement in the positive direction along the $x$ axis. Therefore, the motion of these walls should be examined separately. Let us describe the displacements of the corresponding walls in the form of flat waves

$$\begin{aligned} U_{2n} &= U_0(2n)\cdot \exp(i[kd2n-\omega t]), \\ U_{2n-1} &= U_0(2n-1)\cdot \exp(i[kd(2n-1)-\omega t]), \end{aligned} \tag{4.18}$$

where $U_0(2n)$ and $U_0(2n-1)$ are the amplitudes of the corresponding vibrations, $k$ is the wave vector.

Substituting in turn $U_{2n}$ and $U_{2n-1}$ from (4.18) to the equation of the wall motion (4.4) and cancelling by the exponents corresponding to them, we obtain a set of equations

$$\begin{aligned} &-m\omega^2 U_0(2n) + \tilde{K}U_0(2n) + \\ &\quad +2\sum_{n=1}^{\infty}\tilde{K}(2n-1)\cos kd(2n-1)\cdot U_0(2n-1) + \\ &\qquad +2\sum_{n=1}^{\infty}\tilde{K}(2n)\cos kd2n\cdot U_0(2n) = 0, \\ &-m\omega^2 U_0(2n-1) + \tilde{K}U_0(2n-1) + \\ &\quad +2\sum_{n=1}^{\infty}\tilde{K}(2n-1)\cos kd(2n-1)\cdot U_0(2n) + \\ &\qquad +2\sum_{n=1}^{\infty}\tilde{K}(2n)\cos kd2n\cdot U_0(2n-1) = 0. \end{aligned} \tag{4.19}$$

The latter is the set of homogeneous equations with regard to unknown and undetermined in the linear approximation amplitudes $U_0(2n)$ and $U_0(2n-1)$. Grouping the coefficients of amplitudes $U_0(2n)$ and $U_0(2n-1)$ and equating to zero the determinant consisting of these coefficients, we obtain the equation for determining the unknown frequencies of vibrations

$$\begin{vmatrix} -m\omega^2 + \tilde{K}_1 & \tilde{K}_2 \\ \tilde{K}_2 & -m\omega^2 + \tilde{K}_1 \end{vmatrix} = 0,$$

$$\begin{aligned} \tilde{K}_1 &= \tilde{K} + 2\sum_{n=1}^{\infty}\tilde{K}(2n)\cos 2nkd, \\ \tilde{K}_2 &= 2\sum_{n=1}^{\infty}\tilde{K}(2n-1)\cos(2n-1)kd. \end{aligned} \tag{4.20}$$

This equation has two solutions

$$\omega_-^2=\left(\tilde{K}_1-\tilde{K}_2\right)/\tilde{m},\quad \omega_+^2=\left(\tilde{K}_1+\tilde{K}_2\right)/\tilde{m}, \tag{4.21}$$

which together with taking into account specific expressions for $\widetilde{K}$ and $\widetilde{K}(2n)$, $\widetilde{K}(2n-1)$ can be written in the form

$$\omega_{\mp}^2=\frac{16P_0^2L_y}{\sqrt{\varepsilon_a\varepsilon_c}\,\tilde{m}}\cdot\left\{\sum_{n=1}^{\infty}\ln\left[1+\frac{\varepsilon_a}{\varepsilon_c}\frac{L_z^2}{d^2}\frac{1}{(2n-1)^2}\right]\times\right.$$
$$\times 1\left[1\mp\cos(2n-1)kd\right]-$$
$$\left.-\sum_{n=1}^{\infty}\ln\left[1+\frac{\varepsilon_a}{\varepsilon_c}\frac{L_z^2}{d^2}\frac{1}{(2n)^2}\right]\cdot\left[1-\cos 2nkd\right]\right\}. \tag{4.22}$$

Substituting (4.22) into any of the equations of set (4.19) shows that

$$\left(\frac{U_0(2n)}{U_0(2n-1)}\right)_-=1,\quad \left(\frac{U_0(2n)}{U_0(2n-1)}\right)_+=-1. \tag{4.23}$$

Thus, according to (4.22)–(4.23) there are two branches of vibrations of the domain structure (Fig.7.11). The first branch $\omega_-(\mathbf{k})$ corresponds to vibrations of the acoustic type (they will be referred to as quasi-acoustic). Here $\omega_-(0)=0$, $\omega\sim k$ at low $k$ and the displacement of the adjacent domain walls takes place in one direction. For the second branch

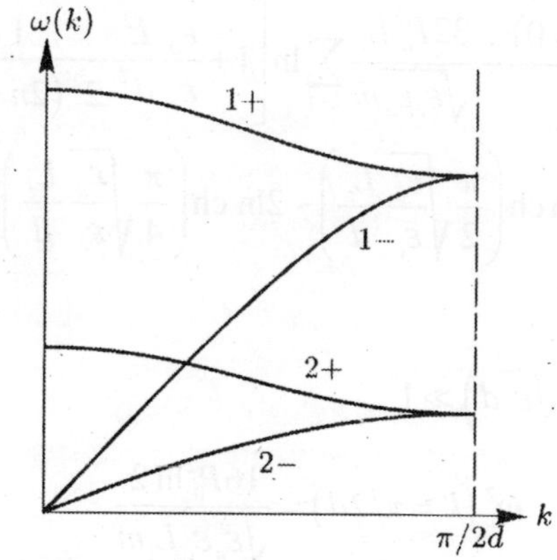


**Fig.7.11.** Frequency dispersion of translational vibrations of the domain structure in quasi-acoustic (–) and quasi-optical (+) branches. 1) without piezoeffect, 2) with the piezoeffect taken into account.

$$\omega_+^2(0)=\frac{32P_0^2L_y}{\sqrt{\varepsilon_c\varepsilon_a}\tilde{m}}\cdot\sum_{n=1}^{\infty}\ln\left[1+\frac{\varepsilon_a}{\varepsilon_c}\frac{L_z^2}{d^2}\frac{1}{(2n-1)^2}\right]=$$
$$=\frac{32P_0^2L_y}{\sqrt{\varepsilon_c\varepsilon_a}\tilde{m}}\ln\prod_{n=0}^{\infty}\left[1+\frac{\varepsilon_a}{\varepsilon_c}\frac{L_z^2}{d^2}\frac{1}{(2n+1)^2}\right]= \qquad (4.24)$$
$$=\frac{32P_0^2L_y}{\sqrt{\varepsilon_c\varepsilon_a}\tilde{m}}\ln\operatorname{ch}\left(\frac{\pi}{2}\sqrt{\frac{\varepsilon_a}{\varepsilon_c}}\frac{L_z}{d}\right).$$

At conventional $\varepsilon_c\sim10^3$, $\varepsilon_a\sim10$, $L_z\sim10^{-1}$ cm, $d\sim10^{-4}\div10^{-3}$ cm $\frac{\pi}{2}\sqrt{\frac{\varepsilon_a}{\varepsilon_c}}\frac{L_z}{d}\gg1$. This yields

$$\omega_+^2(0)\simeq\frac{16\pi P_0^2}{\varepsilon_c md}. \qquad (4.25)$$

Thus, here $\omega_+(0)\neq(0)$, the adjacent walls move in the opposite directions, and the vibrations of this type are naturally referred to as quasi-optical.

At the boundary of the Brillouin zone at $k = \pi/2d$ (the period of the domain structure is $2d$) in (4.22) $\cos(2n-1)kd = 0$, and that is why there is no gap in the spectrum between the optical and acoustic vibrations. Direct calculation of $\omega^2$ yields here

$$\omega_\mp^2(k=\pi/2d)=$$
$$=\frac{\omega_+^2(k=0)}{2}-\frac{32P_0^2L_y}{\sqrt{\varepsilon_c\varepsilon_a}\tilde{m}}\sum_{n=1}^{\infty}\ln\left[1+\frac{\varepsilon_a}{\varepsilon_c}\frac{L_z^2}{d^2}\frac{1}{2^2(2n-1)^2}\right]=$$
$$=\frac{16P_0^2L_y}{\sqrt{\varepsilon_c\varepsilon_a}\tilde{m}}\cdot\left[\ln\operatorname{ch}\left(\frac{\pi}{2}\sqrt{\frac{\varepsilon_a}{\varepsilon_c}}\frac{L_z}{d}\right)-2\ln\operatorname{ch}\left(\frac{\pi}{4}\sqrt{\frac{\varepsilon_a}{\varepsilon_c}}\frac{L_z}{d}\right)\right]. \qquad (4.26)$$

At $\left(\pi\sqrt{\varepsilon_a}L_z/2\sqrt{\varepsilon_c}d\right)\gg1$

$$\omega_\mp^2(k=\pi/2d)\simeq\frac{16P_0^2\ln 2}{\sqrt{\varepsilon_a\varepsilon_c}L_zm}. \qquad (4.27)$$

It is difficult to carry out in the general form the analytical calculations of the spectrum of the domain structure vibrations with the piezoelectric effect taken into account. However, this can be carried out for certain characteristic points of the spectrum. Let us calculate, in particular, the frequency of vibrations of the domain structure in the quasi-optical branch for limiting ($k$=0) vibrations. To calculate the effective mass of the domain wall let us calculate the average value of the kinetic energy of the material of the specimen associated with involvement into motion of its medium elements as the result of the piezoelectric effect. In the limit under consideration ($k$=0) all domain walls are displaced at any moment of time over specific identical distances $U$ and an electric field appears in the entire volume of the specimen, which is practically homogeneous almost everywhere, with the exception of the thin subsurface layer with thickness $d$ (this can be ignored taking into account the ratio $d \ll L$). On the basis of (1.3) and taking into account (4.6) its value in the approximation of weak piezoelectric strains is equal to:

$$E = -\frac{4\pi\bar{\sigma}_1}{\varepsilon_c} = -\frac{4\pi}{\varepsilon_c} \cdot 2P_0 \frac{U}{d}. \tag{4.28}$$

This field results in piezoelectric deformation of the material, the magnitude of which in a static case (here and later to simplify considerations, the tensor nature of the resultant ratios is ignored) on the basis of (1.2), the first of the ratios in (1.3) and ratio (4.28) is as follows

$$\tilde{u} = -\frac{\beta E}{c} = \frac{8\pi P_0}{\varepsilon_c d} \frac{\beta}{c} U. \tag{4.29}$$

This piezoelectric deformation corresponds to the specific displacement of points of the medium whose mean value taking into account the linear form of the ratio of the mentioned values for simple elongation (reduction of thickness) of the specimen, to which we restrict ourselves here, ignoring the simultaneous alteration of the transverse dimensions, is equal to

$$u_3 = L_z \cdot \tilde{u}/2. \tag{4.30}$$

Taking into account that in the case under consideration the entire volume of the material is involved in the motion, and writing the mean kinetic energy of the elements forming it, which is calculated for the unit of the specimen surface in the direction perpendicular to the polar axis

$$\frac{\rho \dot{u}_3^2}{2} \cdot L_x = \frac{m_s^* \omega^2 U^2}{2}, \tag{4.31}$$

we can determine the effective mass of the unit of the specimen area

$$m_s^* = \rho \cdot \frac{(4\pi)^2 P_0^2 \beta^2 L_z^2 L_x}{\varepsilon_c^2 d^2 c^2}. \tag{4.32}$$

Since this mass is distributed between $L_x/d$ walls, for the effective mass of the unit of area of a single wall we have [206, 207]

$$m_{dw}^* = \frac{16\pi^2 P_0^2 \beta^2}{\varepsilon_c^2 \rho c_l^4} \cdot \frac{L_z^2}{d}, \tag{4.33}$$

where $c_l$ is the velocity of the corresponding sound wave.

The estimates of the value $m_{dw}^*$ at conventional $P_0$~$10^4$ (here, as previously, we use CGSE units), $\beta$ ~$10^6$, $L_z$~1, $d$~$10^{-4}$, $\varepsilon_c$~$10^{-3}$, $\rho$~5, $c_l$~$10^5$ CGSE units gives $3 \cdot 10^{-2}$ g·cm$^{-2}$, which is many orders of magnitude higher than the conventional effective mass of the domain walls [143], which is linked with the conversion of the spontaneous polarization in the region of the moving domain wall and which for the same values of the constants included in this equation is equal to ~$10^{-11}$ g/cm$^2$.

Evidently, the obtained increase of the effective mass of the domain wall also greatly reduces the frequency of natural vibrations of the domain walls (Fig.7.11). In particular, according to (4.25) for limiting long-wave quasi-optical vibrations instead of the value ~$10^{10}$ Hz without piezoelectric effect taken into account, the value of the corresponding frequency with the piezoelectric effect taken into account decreases to ~$10^6$ Hz, i.e. to megahertz frequencies.

For the same reasons as in the case of ferroelectrics, translational vibrations of the domain walls also occur in the ferroelastics. For arbitrary values of the wave vector such vibrations have not been studied but, for the case of $k$=0, the frequency of the corresponding vibrations can be easily estimated, avoiding labour-consuming calculations. For this purpose, let us first of all calculate the increase of the elastic energy of the ferroelastic associated with the equal displacements of the domain walls. The average deformation in the material of each domain at displacement of the domain boundaries by the value $U$ is equal to $\tilde{u} = 2\epsilon_0 U/d$. Then, for a contact with an absolutely rigid material, where all elastic fields are concentrated in the material of the ferroelastic, this increase of elastic energy for a single domain is

$$\Phi = \frac{1}{2}\mu \tilde{u}^2 V = \frac{1}{2}\mu \epsilon_0^2 U^2 L_y L_z / d. \tag{4.34}$$

Relating these values of $\Phi$ to the unit of area of the domain wall and separating here the coefficient of $U^2/2$, we determine the coefficient of the quasi-elastic force acting on the unit of area of the domain wall

$$K = \tilde{K}/L_y L_z = \frac{4\mu\epsilon_0^2}{d}, \tag{4.35}$$

where, as above, $\mu$ is the elastic module and $\epsilon_0$ is the magnitude of spontaneous deformation.

Taking into account that the average displacement of the elements of the material of the ferroelastic at displacement of the domain boundary by the value $U$ is (the direction of spontaneous shear coincides with axis $y$)

$$u_2 = \frac{\epsilon_0 L_z}{2} \frac{U}{d}, \tag{4.36}$$

we can determine the kinetic energy of the material of a single domain

$$\frac{\rho \dot{u}_2^2}{2} \cdot L_y L_z d = \frac{\rho}{2} \frac{\epsilon_0^2 L_x^2}{4d} \omega^2 U^2 L_y L_z = \frac{m_{dw}^* \omega^2 U^2}{2} L_y L_z. \tag{4.37}$$

Hence, the effective mass of the unit of area of the domain wall is [209–212]

$$m_{dw}^* = \frac{\rho \varepsilon_0^2 L_x^2}{4d}. \tag{4.38}$$

Taking into account (4.35) and (4.38) the frequency of vibrations of the ferroelastic domain boundaries in the quasi-optical branch at $k$=0 is

$$\omega_+ (k=0) = \sqrt{\frac{K}{m_{dw}^*}} = \frac{\pi}{L_x} \sqrt{\frac{\mu}{\rho}}. \tag{4.39}$$

As it was expected, this frequency coincides with the resonance frequency of elastic shear vibrations of the ferroelastic plate in respect of thickness. For other values of $k$ there will be no such coincidence of course.

## 7.5. NATURAL AND FORCED TRANSLATIONAL VIBRATIONS OF DOMAIN BOUNDARIES IN REAL FERROELECTRICS AND FERROELECTRICS – FERROELASTICS

The natural undamped translational vibrations of domain boundaries in ferroelectrics and ferroelastics with defects are described in a simple manner within the framework of the accepted approach. As in the case of bending vibrations of the domain walls, in this case the quasi-elastic term $\overline{K}\,U_n$ is added to the equation of the boundary motion with the same coefficient $\overline{K}$ as in section 7.3, which characterizes the interaction of the boundary with the system of defects. This will result in the increase of rigidity of the domain structure in relation to the translational displacement of the boundaries and, consequently, in the increase of all frequencies of vibrations in the both branches. A completely new issue here is the absence of vanishing of the dependence $\omega(\mathbf{k})$ in the initial quasi-acoustic branch at $k \to 0$ in a real material [213].

There are more options of practical importance when damped vibrations are considered particularly in those cases when the damping is caused by the relaxation of domain walls interacting with the defects. Let us investigate special features of such a motion of the domain walls using the example of their interaction with a system of point defects.

Depending on the mobility of defects, their influence on the mobility of domain boundaries can be implemented in two qualitatively different ways – by the forces of dry or viscose friction, respectively. In the first case, the motion of the domain wall is the motion through a system of stationary obstacles consisting of successive acts of detachment of the boundary from stationary stoppers with its further capture by other defects. The second case takes place if the domain wall interacts with a system of mobile defects accompanying its motion [194,195].

In accordance with the concepts of chapter 5 and section 7.3, both types of motion can be described using the one-dimensional model. At the same time, regardless of the general nature of consideration, because of the difference in terminology it is convenient to carry out their specific description separately. Let us consider the first of these types of motion when the domain wall moves in the external field through a system of stationary stoppers. At that it is taken into account that in accordance with the results in chapter 5 the domain boundary in a real crystal has a deformed profile already in the initial equilibrium position being captured by the neighbouring defects.

Let us determine the expression for the force acting on the moving boundary from the direction of the defects interacting with the boundary. The power per unit of area of the domain wall, required by the external electric field to overcome the resistance of defects to displacement of the arbitrary domain wall, is equal to

$$F \cdot \dot{U} = \int \frac{\partial \mathcal{U}}{\partial t} n(x,t)\,dx =$$
$$= \int \frac{\partial \mathcal{U}}{\partial U} \cdot \dot{U} \cdot n(x,t)\,dx = -\dot{U} \int \frac{\partial \mathcal{U}}{\partial x} \cdot n(x,t)\,dx. \tag{5.1}$$

Here $\mathcal{U}=\mathcal{U}(x-U(t))$ is the increase of the energy associated with the bending of the wall from its equilibrium position (symmetric in this case) in the system of points of its pinning by defects, i.e. finally, linked with extra bending of the domain wall, $U$ is the coordinate of the plane of the average orientation of the domain wall interacting with the defects, $\dot{U}$ is its velocity, $n(x,t)$ is the volume concentration of points of the boundary pinning by the defects.

The time dependence of the distribution of pinning points $n(x,t)$ is described by kinetic equation with the single relaxation time:

$$\frac{dn}{dt} = -\frac{n-n_{\infty}}{\tau}, \tag{5.2}$$

where $n_{\infty}(x)$ is the equilibrium distribution of pinning points in the given region of the crystal, which in accordance with the results of chapter 5, can be regarded as having a stepped form:

$$n_{\infty} = n \cdot \Theta\left(\tilde{U} - |x|\right). \tag{5.3}$$

Here $\Theta(x)$ is Heaviside's function, $n$ is the volume concentration of the defects, displacement $\tilde{U}$, as previously, characterizes the maximum distance of the defect from the plane of the average orientation of the boundary at which the boundary is still captured by the defect.

The solution of (5.2) has the form

$$n(x,t) = \int_{-\infty}^{t} \exp\left(-\frac{t-\xi}{\tau}\right) n_{\infty}(\xi) \frac{d\xi}{\tau}. \tag{5.4}$$

Substituting (5.4) into the expression for the pressure of the force acting on the boundary from the direction of the defects, which according to (5.1) is

$$F = -\int \frac{\partial \mathcal{U}}{\partial x} n(x,t)\,dx \tag{5.5}$$

and replacing here the difference $x-U(t)$ by $x$ we obtain the following [214]:

$$F=-\int\frac{\partial\mathcal{U}(x-U(t))}{\partial x}\int_{-\infty}^{t}\exp\left(-\frac{t-\xi}{\tau}\right)n_{\infty}(x-U(\xi))\frac{d\xi}{\tau}dx=$$
$$=\int_{-\infty}^{t}\exp\left(-\frac{t-\xi}{\tau}\right)\int_{-\infty}^{\infty}\frac{\partial\mathcal{U}(x)}{\partial x}n_{\infty}\left[U(t)-U(\xi)\right]dx\frac{d\xi}{\tau}=$$
$$=\int_{-\infty}^{t}\exp\left(-\frac{t-\xi}{\tau}\right)\mathcal{E}\left[U(t)-U(\xi)\right]\frac{d\xi}{\tau},\tag{5.6}$$

where

$$\mathcal{E}\left[U(t)-U(\xi)\right]=\int_{-\infty}^{\infty}\frac{\partial\mathcal{U}(x)}{\partial x}n_{\infty}\left[U(t)-U(\xi)\right]dx.\tag{5.7}$$

When determining the motion of the domain boundaries in harmonic approximation, the difference $\Delta\equiv U(t)-U(\xi)$ will be regarded as small. Then taking into account the specific type of $n_{\infty}(x)$ (5.3) we have

$$\varepsilon\left[U(t)-U(\xi)\right]\equiv\varepsilon[\Delta]=\varepsilon(0)+\frac{\partial\varepsilon}{\partial\Delta}\Big|_{0}\cdot\Delta+\ldots\simeq$$
$$\simeq n\vartheta\tilde{U}\left[U(t)-U(\xi)\right]\equiv K\left[U(t)-U(\xi)\right],\tag{5.8}$$

where $\vartheta$ and $\tilde{U}$ for a pure ferroelectric are determined by the expressions from chapter 5, and for a ferroelectric–ferroelastic by the equations from this chapter, respectively.

Let us further consider the motion of the domain boundaries in the external field. Since the strength of the later depends only on time and is almost independent of the coordinates, the motion of all domain walls in this field will be identical. Therefore, when writing equations of motion of an arbitrary wall, the number of the displaced wall can be omitted and, taking this into account, the mentioned equation can be written in the form

$$m^{*}\ddot{U}+KU(t)+F[U]=2P_{0}E(t),\tag{5.9}$$

where $K=32P_0^2\ln ch\left(\frac{\pi}{2}\sqrt{\frac{\varepsilon_a}{\varepsilon_c}}\frac{L_z}{d}\right)\Big/\sqrt{\varepsilon_c\varepsilon_a}L_z$ is the coefficient of the quasi-elasticity of the domain wall in a defect-free crystal related to the unit of area of the wall, which evidently is equal to the coefficient of quasi-elasticity in the optical branch of the vibrations, taken at $k$ = 0 (4.24), force $F[U]$ is specified by expression (5.6), and term $2P_0E$ describes the pressure on the domain wall in the external field.

Replacing $F[U]$ in (5.9) by the expression (5.6) where $\mathcal{E}[U(t)=U(\xi)]$ is determined by expression (5.8), we obtain

$$m^{*}\ddot{U}+K\cdot U(t)+\overline{K}U(t)-\overline{K}\int_{-\infty}^{t}\exp\left(-\frac{t-\xi}{\tau}\right)\cdot U(\xi)\frac{d\xi}{\tau}=2P_0E(t). \quad (5.10)$$

Let the external electric field change with time in accordance with the harmonic law $E=E_0\exp(i\omega t)$. Let us look for the solution of equation (5.10) for steady motion in the form of $U(t)=U_0\exp(i(\omega t+\alpha))$. Substituting it into (5.10) in general case we obtain [194, 195]:

$$U(t)=\frac{2P_0E(t)}{\left[-m^{*}\omega^{2}+\left(K+\overline{K}\right)-\overline{K}/(1+i\omega\tau)\right]}. \quad (5.11)$$

In the practically important case of relatively low frequencies $\omega$ when the inertial term in (5.4) and, consequently, in (5.11) can be ignored, we have

$$U(t)=\frac{2P_0E_0}{\left(K+\overline{K}\right)}\times$$

$$\times\left\{\left[1+\frac{\overline{K}/K}{\left(1+i\omega^{2}\tau_c^{2}\right)}\right]\cos\omega t+\frac{\omega\tau_c\overline{K}/K}{\left(1+i\omega^{2}\tau_c^{2}\right)}\sin\omega t\right\}, \quad (5.12)$$

where

$$\tau_c=\tau\cdot\frac{\left(K+\overline{K}\right)}{K}. \quad (5.13)$$

The motion of the domain wall, interacting with a system of mobile defects, is described using the above approach, with the only difference that the role of $n(x,t)$ in all expressions here is performed directly by the concentration of defects in the given location in the crystal and therefore $\tau$ here is the relaxation time of the defective atmosphere. Thus, in contrast to previous discussion, devoted to consideration of the passage of the domain wall through the system of stationary stoppers, where time $\tau$ characterizes the relaxation properties of the domain wall, determined primarily by the energy of its interaction with defects, here time $\tau$ characterizes the mobility of defects, i.e. depends mainly on the activation energy of its motion.

When the crystal contains relatively mobile defects, the domain wall in the initial states is flat, but enriched (or depleted, depending on the sign of energy $\mathcal{U}_0$ of the boundary interaction with the defects) by the atmosphere of the defects. Substitution in (5.7) in the case of interaction of the boundary with the atmosphere of mobile defects of

$$n_\infty(x)=n+n\left(\exp(\mathcal{U}_0/T)-1\right)a\delta\left(x-U(t)\right), \tag{5.14}$$

as the equilibrium distribution [215] shows that the equation of motion of the domain wall and its solution here are also described by the expressions (5.10)–(5.13), in which $\overline{K}=n\vartheta\widetilde{U}$ is replaced by $\overline{K}=na\left(\exp(\mathcal{U}_0/T)-1\right)\vartheta$.

The difference in the description of the cases of ferroelectrics and ferroelectrics–ferroplastics for both types of motion of the domain walls consists of application of different coefficients of $\vartheta$.

## 7.6. DOMAIN CONTRIBUTION TO THE INITIAL DIELECTRIC PERMITTIVITY OF FERROELECTRICS. DISPERSION OF THE DIELECTRIC PERMITTIVITY OF DOMAIN ORIGIN

The domain boundaries carry out repolarization of the ferroelectric material by displacing in the external fields and, consequently, contribute to its dielectric permittivity $\varepsilon$ [216–231]. The magnitude of this contribution taking into account the definition of the dielectric constant is evidently

$$\varepsilon=\frac{4\pi\Delta P}{E}=\frac{8\pi P_0}{E}\cdot\frac{U}{d}. \tag{6.1}$$

Substituting the expression for $U$(5.12) into (6.1) shows that in this case it has the usual Debye form

$$\varepsilon(\omega)=\varepsilon_\infty+\frac{\left(\varepsilon_0-\varepsilon_\infty\right)}{1+i\omega\tau_c}, \tag{6.2}$$

the real and imaginary parts of which are

$$\varepsilon'=\varepsilon_\infty+\frac{\left(\varepsilon_0-\varepsilon_\infty\right)}{\left(1+i\omega^2\tau_c^2\right)},\quad \varepsilon''=\frac{\left(\varepsilon_0-\varepsilon_\infty\right)\omega\tau_c}{1+i\omega^2\tau_c^2}. \tag{6.3}$$

Here $\varepsilon_0$ and $\varepsilon_\infty$ are the static (i.e. measured at $\omega=0$) and high-frequency ($\omega\to\infty$) dielectric permittivities respectively, which in accordance with the results of sections 7.4 and 7.5 are controlled by the values of $K$ and $\overline{K}$, i.e. by the charges on the surface of the ferroelectric material or the elastic influence on the displacements of the boundaries of the surface non-ferroelectric layer and by interaction of the boundaries with the defects. In particular, in the crystal of the ferroelectric–ferroelastic with stationary defects, ignoring the elastic contribution to the value of $K$

$$\varepsilon_0 = \frac{16\pi P_0^2}{dK} = \frac{\pi L_z \sqrt{\varepsilon_c \varepsilon_a}}{2d \ln \mathrm{ch}\left(\pi \sqrt{\varepsilon_a}\, L_z / 2\sqrt{\varepsilon_c} d\right)}, \tag{6.4}$$

$$\varepsilon_\infty = \frac{16\pi P_0^2}{d\left(K+\overline{K}\right)} \simeq \frac{4\sqrt{2}}{nd}\left(\frac{\pi^3 \tilde{\varepsilon} P_0^3}{\mu \epsilon_0^2 \mathcal{U}_0^2 a^2}\right)^{1/4}. \tag{6.5}$$

In the crystal of a pure ferroelectric within the framework of the previously mentioned approximation, the value of $\varepsilon_0$ remains the same, and the value of $\varepsilon_\infty$ becomes the following

$$\varepsilon_\infty = \frac{8P_0\sqrt{P_0}\tilde{\varepsilon}^{1/4}}{(\gamma a)^{1/4}\, n\mathcal{U}_0^{1/2} d}. \tag{6.6}$$

In both cases

$$\varepsilon_0 - \varepsilon_\infty = \frac{16\pi P_0^2 \overline{K}}{dK\left(K+\overline{K}\right)}. \tag{6.7}$$

The height of the maximum is

$$(\mathrm{tg}\,\delta)_{\max} = \frac{(\varepsilon_0 - \varepsilon_\infty)}{2\sqrt{\varepsilon_0 \varepsilon_\infty}} = \frac{1}{2}\frac{nk\tilde{U}}{\sqrt{K\left(K+n\vartheta\tilde{U}\right)}}. \tag{6.8}$$

If the coefficient of the quasi-elastic force, acting on the boundary from the direction of the defects (for determinacy, the case of the ferroelectric–ferroplastic is selected),

$$\overline{K} \equiv n\vartheta\tilde{U} = 2\sqrt{2}\left(\pi\mu/\tilde{\varepsilon}\right)^{1/4} n\sqrt{P_0 \epsilon_0 \mathcal{U}_0 a} \tag{6.9}$$

is considerably greater than $K$, the height of the maximum of $\mathrm{tg}\,\delta$ is

$$(\mathrm{tg}\,\delta)_{max} \simeq \frac{1}{2}\sqrt{\frac{n\vartheta\tilde{U}}{K}} = \frac{1}{4}\sqrt{\left(\frac{\mu\tilde{\varepsilon}^3}{\pi^3}\right)^{1/4} \frac{nd\sqrt{\mathcal{U}_0 \epsilon_0 a}}{P_0\sqrt{P_0}}}. \tag{6.10}$$

Equations (6.8) and (6.10) show that with the increase of the concentration of defects, the initially linear growth of the height of the maximum is subsequently replaced by a root dependence, i.e.

the rate of growth of the height of the tg$\delta$ maximum gradually decreases with the increasing $n$.

The general expression for the components of dielectric permittivity and the associated values, written using the quasi-elastic constants $K$ and $\overline{K}$, is evidently preserved in the case of the atmosphere of mobile defects as well. In this case, since constant $\overline{K}$ is proportional to the concentration of defects $n$, regardless of whether they are mobile or not, the concentration dependence of the components of dielectric permittivity for mobile defects in comparison with the case of the domain boundary passage through a system of stationary stoppers remains unchanged, whereas their temperature dependences are expected to be more sharp in this case.

The high values of dielectric permittivity represent one of the main distinguishing features of the ferroelectrics. To some extent they are linked with the anomalous softness of the crystalline lattice in the vicinity of the Curie point, and partially with the contribution to the value of $\varepsilon$ of the domain boundaries displacement in the external field.

Let us compare the obtained theoretical dependences with the experimental data. The presence of a domain contribution to the values of dielectric permittivity was found experimentally for almost all main groups of the ferroelectric crystals – TGS, $BaTiO_3$, $KH_2PO_4$ etc. [232–241]. In the case of crystals of the family of potassium dihydrophosphate this contribution is especially large, very distinctive and most thoroughly studied. The domain nature of anomalously high values of dielectric permittivity in the region of the so-called 'plateau' is confirmed here in a large number of experiments. Among them are the decrease of the values of dielectric permittivity down to the transition of the dependence $\varepsilon(T)$ to the corresponding dependence for a monodomain crystal when an external electric field of sufficiently high strength (Fig.7.12) is applied to the crystal [232], and a decrease of the values of $\varepsilon$ when defects are purposefully introduced into a crystal during doping of the crystal in the process of its growth (Fig.7.13) [233], exposure of the crystal to radiation or particles of various types [242–245] and, finally, visual observation of the domain structure of the crystal located in an external field [183].

The experimentally observed domain contribution to $\varepsilon$ in the nominally pure crystals can be described by the expression for $\varepsilon_0$. The experiment shows for the KDP crystals that the value of $\varepsilon$ here, as well as the constant $\varepsilon_0$ in (6.4), depends on the thickness

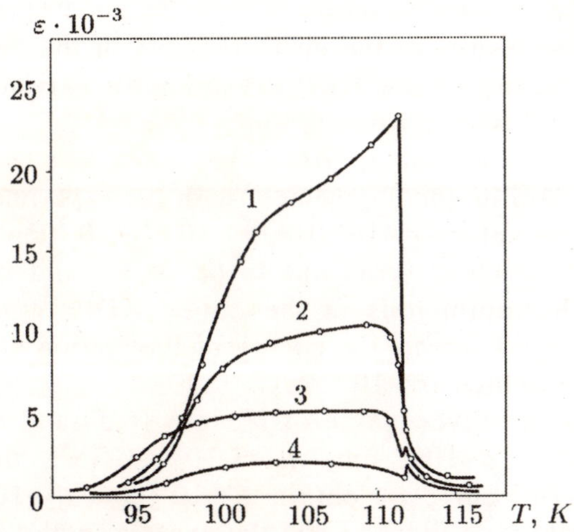


**Fig.7.12.** Dependence $\varepsilon(T)$ for a RDA crystal measured after applying of a constant field $E$ to the crystal: 1, 2, 3, 4 – $E = 0$; 1; 1.5; 2 kV/cm; $E_{\sim}$=1 V/cm.

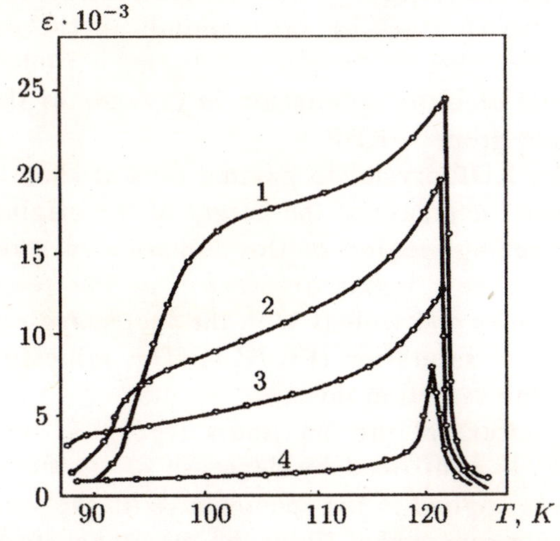


**Fig.7.13.** Dependence $\varepsilon(T)$ for a KDP crystal with different content of chromium ions [233]: 1) nominally pure KDP: 2,3,4 – $n=10^{18}$, $10^{19}$, $10^{20}$ $cm^{-3}$ ; $E_{\sim}$=1 V/cm.

of the ferroelectric plate, and increases in particular with the increasing $L_z$.

The defects appearing in the KDP crystal during doping are evidently stationary. Therefore, the dielectric permittivity of such a material can be described by the expression for $\varepsilon_\infty$. As show in Fig.7.13, the inversely proportional dependence of $\varepsilon_\infty$ on $n$ expected here in accordance with equation (6.5) is in good agreement with the experimental results of the measurements of the values of $\varepsilon$ in the 'plateau' region in the KDP crystal with various degree of doping the crystal with chromium ions.

The numerical estimates of $\varepsilon$ in the cases having been considered are also in good agreement with the experimental values. For conventional values $\varepsilon_c \sim 10^3 - 10^4$, $\varepsilon_a \sim 10$, $L_z \sim 10^{-1}$ cm, $d \sim 10^{-4}$ cm, the constant $K$ in (6.4) turns out to be $\sim 10^{10}$, and the value of $\varepsilon_0 \sim 10^4$. The chromium ions in the doped KDP crystals can be regarded as charged defects the energy of interaction of the domain boundaries with which is $\sim 10^{-13}$ erg.

Substituting the given value of $\mathcal{U}_0$ into (6.5) and also $\tilde{\varepsilon} \sim 10^2$, $P_0 \sim 10^4$, $\epsilon_0 \sim 10^{-2}$, $d \sim 10^{-4}$ cm, $\mu_a \sim 3 \cdot 10^{10}$ CGSE units [183], $a \sim 10^{-7}$ cm, $n \sim 10^{18}$ cm$^{-3}$, we obtain $\overline{K} \sim 10^{11}$ and $\varepsilon_\infty \sim 10^3$, which is also in agreement with Fig.7.13 and the experimental data published in [232].

The general form of the dependences of the components of dielectric permittivity, $(\text{tg}\delta)_{max}$ on the concentration of defects, specified by ratio (6.2)–(6.8), corresponds to the experimental results of the interaction of domain boundaries with defects formed under gamma and electron irradiation in crystals of the potassium dihydrophosphate group – KDP.

Exposure of a KDP crystal to gamma rays at room temperature results in a gradual decrease of the height of the original maximum $(\text{tg}\delta)_I$ in the 'freezing' region of the domain structure and in the appearance of a new maximum $(\text{tg}\delta)_{II}$ at the temperature of ~108 K, which grows and widens with the increasing radiation dose $D$ and then tends to saturation (Fig7.14). The values of $\varepsilon$ decrease with the increasing radiation dose.

The domain nature of the maximum $(\text{tg}\delta)_{II}$, as well as of the maximum $(\text{tg}\delta)_I$, is confirmed by decrease of its magnitude when a constant field is applied to the specimen. With the variation of the frequency of the measurement field, the maximum $(\text{tg}\delta)_{II}$ is shifted along the temperature scale, which indicates its relaxation character.

The influence of gamma irradiation on the dielectric properties of the crystals of the KDP group at room temperature are confirmed by irradiation of crystals at the temperature of liquid nitrogen. It is important that immediately after irradiation at the liquid nitrogen temperature peak $(\text{tg}\delta)_{II}$ is absent and appears only during annealing. It achieves the maximum magnitude after annealing at ~213 K, and then decreases in magnitude and completely vanishes after annealing at ~293 K [242, 243].

The results of the influence of exposure to the electron irradiation on the dielectric properties of the $KH_2PO_4$ and $CsH_2AsO_4$ crystals at room temperature are qualitatively similar to the case of gamma radiation.

As mentioned above, the concentration dependence of the components of dielectric permittivity at different types of motion of the domain boundaries in crystals with defects (motion of the domain walls through the system of stationary stoppers or motion that involves dragging defect atmosphere) is the same. In such conditions, the answer to the question: which type of motion of the domain walls occurs in this experiment? – is given by the temperature dependences of $\varepsilon$ and tg$\delta$.

Comparison of Figs. 7.13 and 7.14 shows that the dependences $\varepsilon$ ($T$) in the latter case are more 'distinctive'. In addition to this, at the similar concentration of defects the behaviour of tg$\delta$ ($T$) differs greatly here (in the case of the KDP crystal with chromium the new maximum of tg$\delta$ does not appear). All these factors indicate that the mechanism of motion of the domain boundaries in the doped and $\gamma$, $e^-$ irradiated crystals differs qualitatively. And since the doped crystals are unambiguously characterized by the occurrence of the mechanism of the boundary motion through a system of stationary stoppers, then in the irradiated crystals the domain boundaries motion of the viscose friction type is of the highest probability.

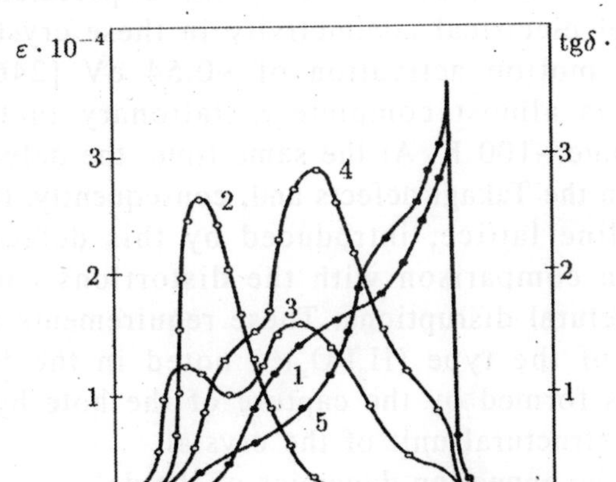


**Fig.7.14.** Temperature dependences $\varepsilon$ (1,5) and tg $\delta$ (2,3,4) of a KDP crystal prior to and after electron irradiation ($E_{\sim}$=1V/cm, $f$=1 kHz). 1,2 – $nvt$ (integral flux) = 0; 3 – $nvt$ ~$5\cdot10^{14}$ e·cm$^{-2}$; 4,5) – $nvt$~$5\cdot10^{15}$ e·cm$^{-2}$; ($nv\simeq10^{12}$ e·cm$^{-2}$·s$^{-1}$).

The application of the condition $\omega\tau_c = \sqrt{\varepsilon_\infty/\varepsilon_0}$ of the maximum tg$\delta$ observation and of the ratio (5.13) that links the relaxation time of the domain boundaries in this case with the relaxation time of defects $\tau$ enables us to determine the latter at the temperature of tg$\delta$ maximum. On the basis of the above ratio and the relation $\varepsilon_\infty/\varepsilon_0=K/(K+\overline{K})$ we have

$$\tau_{max} = \frac{1}{\omega}\left(\frac{\varepsilon_\infty}{\varepsilon_0}\right)^{3/2}. \tag{6.11}$$

The effect of irradiation is revealed most distinctively starting with the integral flux of $nvt$~$5\cdot10^{15}$ e·cm$^{-2}$. The ratio of the dielectric permittivities prior to and after irradiation at the temperature of tg$\delta$ maximum here (Fig.7.14) is equal to approximately 2–3. Substituting into (6.11) the mentioned ratio and $\omega=2\pi\cdot10^3$, we obtain $\tau_{max}$~$(2-3)\cdot10^{-4}$ s$^{-1}$. Assuming that the temperature dependence of $\tau$ is governed by the conventional activation law $\tau = \tau_0\cdot\exp(\mathcal{U}_0/T)$, at the previously found value of $\tau_{max}$ and the maximum temperature ~108 K, $\tau_0$~$10^{-13}$ s$^{-1}$ we obtain the activation energy of the defect interacting with the boundary $\mathcal{U}_0$~0.2 eV.

The obtained value of activation energy allows us to make a number of assumptions on the nature of the defect interacting with the boundary. This defect should be much more mobile than the proton vacancy which, as shown by the experiments with the measurement of electrical conductivity in these crystals, has the energy of the motion activation of ~0.54 eV [246–248] and, consequently, is almost completely stationary in the relevant temperature range ~100 K. At the same time, the defect should be less mobile than the Takagi defects and, consequently, the distortion of the crystalline lattice, introduced by this defect should be intermediate in comparison with the distortions caused by the mentioned structural disruptions. These requirements are satisfied by the defect of the type $(H_2PO_4)^{2-}$, noted in the EPR spectra [249], which is formed by the capture of the hole by the proton depleted $PO_4^{3-}$ structural unit of the crystal.

Analysis of the annealing dynamics of the defects formed in the process of $\gamma$, $e^-$ irradiation shows that the low temperature (liquid nitrogen temperature) and high temperature (room temperature) irradiation results in the formation of at least qualitatively different types of defects in the KDP group crystals. The low temperature irradiation causes the formation of defects with relative low mobility in the crystal, which result in pinning of the domain boundaries by them, and the high temperature irradiation leads to the creation of mobile defects the interaction of which with the domain boundaries is of the viscose friction type. The defects of the first type are unstable in relation to the temperature [249] and at ~193 K convert to the defects of the second type. The change of the of the crystal colour at low temperature irradiation (light purple colour) indicates that the defects of the first type can be the radicals [250], which were observed in the EPR spectra [249].

The mentioned transformation of the defects in the $\gamma$, $e^-$-irradiated crystals with the change of temperature can be described as follows. Exposure of a crystal to a flux of $\gamma$-beams or an electron beam at relatively high temperatures of the order of room temperature, leads on the one hand to ionisation and on the other to local heating of the crystal as a result of which four protons on the hydrogen bonds adjacent to the given tetrahedron can simultaneously approach it and form a configuration, which is unsuitable for usual conditions. The capture by the resultant configuration of an electron leads to its stabilization and the formation of a relatively stable complex ($H_4PO_4$). With the decrease of temperature in the vicinity of $T$~193 K the tendency of protons to ordering on hydrogen bonds makes two protons leave the complex ($H_4PO_4$) for the other adjacent tetrahedrals, which results in the structural change of this complex and appearance of a relatively mobile defect $(HPO_4)^{2-}$.

Irradiation of the KDP crystals at the liquid nitrogen temperature results in the formation of proton vacancies and double protons on the bond whose energy can be partially reduced by the loss or capture of an electron leading to the formation of $(HPO_4)^-$ and $(H_3PO_4)^+$ structural units of the crystal. Regardless of a partial decrease of the distortions of the lattice around the defect as a result of the loss (capture) of an electron, the units $(HPO_4)^-$ and $(H_3PO_4)^+$ remain low mobile at temperatures below ~193 K and their interaction with the domain boundaries causes the latter to be eliminated from the repolarization processes. With the increase of temperature up to approximately ~193 K the complexes $(HPO_4)^-$ and $(H_3PO_4)^+$ transform to relatively mobile units $(H_2PO_4)^{2-}$, whose interaction with the domain boundary is of the viscose friction nature.

Ratios (6.2)–(6.8) not only correctly describe the main qualitative rules of the influence of defects formed by irradiation, on the dielectric properties of the ferroelectrics of the KDP type (for example, the decrease of the growth rate of the height of the $(\mathrm{tg}\delta)_{II}$ maximum at a specific defect concentration, etc.), but also show good quantitative agreement with the experiments, which, for example, is efficiently demonstrated by the estimates of the height of the $(\mathrm{tg}\delta)_{II}$ maximum depending on the value of $n$.

The dielectric properties of the KDP-type crystals are influenced far more strongly by the exposure to fluxes of fast neutrons [244,245] in comparison with the $\gamma$, $e^{-}$-irradiation. As show in [245], exposure of the investigated crystals to the flux of fast neutrons with $nvt \sim 10^{17}-10^{18}$ neutron·cm$^{-2}$ plus to accompanying gamma irradiation with the dose of $10^{9}-10^{10}$ P results in a drastic change of the nature of dependences $\varepsilon(T)$ and $\mathrm{tg}\delta(T)$: a 'plateau' of dependence $\varepsilon(T)$ is suppressed, $(\mathrm{tg}\delta)_{I}$ disappears and a new maximum is formed at a higher temperature, which evidently has the same origin as $(\mathrm{tg}\delta)_{II}$ in the $\gamma$-irradiated crystals.

The defects formed in the crystals in the case of the neutron irradiation are apparently associated with disruptions not only in the electronic subsystem but also with displacements of heavier ions. Such defects in the conditions of the experimental investigations can be regarded as almost stationary and their interaction with domain boundaries is relatively strong. To describe them, it is necessary to assume $\tau_c \to \infty$ in (6.2)–(6.8), whence in complete agreement with experiments, we obtain $\varepsilon' = \varepsilon_\infty < \varepsilon_0$, $\varepsilon''$, $\mathrm{tg}\delta = 0$ (the $(\mathrm{tg}\delta)_{II}$ maximum in this case is evidently caused by the effect of accompanying $\gamma$-radiation).

The experimental investigations show that the influence of gamma and $x$-radiation on the dielectric properties of deuterated crystals of the KDP group is similar to their influence on the dielectric properties of non-deuterated crystals. It allows to assume that the defects, formed by similar types of irradiation in these crystals are qualitatively identical.

## 7.7. DOMAIN CONTRIBUTION TO THE ELASTIC COMPLIANCE OF FERROELASTICS

A consideration similar to the above also makes it possible to estimate the domain contribution to the values of elastic compliance of ferroelastic materials. In particular, for an isotropic ferroelastic, the effect of elastic stress $\sigma$ conjugate to spontaneous deformation $\epsilon_0$ gives

$$U_{\max} = \tilde{U} = \frac{\sigma \cdot l^2}{2\pi\mu\epsilon_0 a}. \tag{7.1}$$

Then, the domain contribution to the values of the coefficient of elastic compliance is

$$s_{66}^{E}=\frac{\epsilon_0}{\sigma}\frac{\overline{U}}{d}=\frac{l^2}{2\pi\mu a d}. \tag{7.2}$$

The coefficient of the quasi-elastic force, acting on the boundary from the direction of a single defect is

$$\vartheta=4\mu\pi\epsilon_0^2 a. \tag{7.3}$$

Taking into account (7.3) on the basis of the condition $\vartheta\widetilde{U}^2/2=\mathcal{U}_0$

$$\tilde{U}=\left(\frac{\mathcal{U}_0}{2\pi\mu\epsilon_0^2 a}\right)^{1/2} \tag{7.4}$$

and, consequently, the final expression for the coefficient of the elastic compliance is

$$s_{66}^{E}=\frac{\epsilon_0}{\sqrt{2\pi\mu a\mathcal{U}_0 n d}}. \tag{7.5}$$

At $\epsilon_0\sim10^{-2}$, $a\sim10^{-7}$ cm, $\mathcal{U}_0\sim10^{-13}$, $d\sim10^{-4}$, $n\sim10^{-18}$, $\mu\sim3\cdot10^{10}$ the value of $s_{66}^{E}$ turns out to be $\sim3\cdot10^{10}$, i.e. comparable with the elastic constant of the monodomain crystal.

## 7.8. NON-LINEAR DIELECTRIC PROPERTIES OF FERROELECTRICS, ASSOCIATED WITH THE MOTION OF DOMAIN BOUNDARIES

In the previous paragraphs we obtained the expressions for the so-called initial dielectric permittivity, and, correspondingly, initial elastic compliance. The calculations were performed for the case of small deviations of domain boundaries from the equilibrium state in the systems of points of boundaries pinning by the defects. Consequently, the average displacement of the domain boundaries proved to depend linearly on the external field, and the constants of proportionality between them, which with the accuracy to the coefficient corresponds to dielectric constant, turned out to be completely independent of the external field. Such behaviour of the dielectric constant is characteristic of linear dielectrics. At the same time, in ferroelectrics, which are non-linear dielectrics, as it was seen already from considerations in chapter 6, the presence of relatively strong field results in a non-linear dependence of polarization on the external field. In the model of motion of the domain boundaries, interacting with defects, the mentioned behaviour of dielectric permittivity corresponds to the case of relatively large displacements of the domain walls. In fact, in relatively strong fields, a boundary evidently detaches itself from the defects which will cause a sharp increase of the domain wall displacement, and, consequently, the value of the dielectric constant.

The calculation of pressure $F[U]$ per unit of area of the boundary from the direction of the defects in the general case according to (5.5) yields the dependence

$$F[U]=-\int\frac{\partial\mathcal{U}}{\partial x}n(x,t)\,dx=\frac{n\vartheta}{2}\left[2\tilde{U}U-U^2\right], \tag{8.1}$$

which passes through the maximum at

$$F_{max}=\frac{n\vartheta}{2}\tilde{U}^2=n\mathcal{U}_0. \tag{8.2}$$

Equating this pressure to the pressure on the boundary from the direction of the external electric field, we obtain the value of the critical (threshold) field at which the domain boundaries detach from the defects:

$$E_c=\frac{n\mathcal{U}_0}{2P_0}. \tag{8.3}$$

As indicated by (8.3), this field linearly depends on $n$. In addition to this, due to the stronger temperature dependence of the energy of interaction of the boundaries with defects $\mathcal{U}_0$ in comparison with the temperature dependence of spontaneous polarization $P_0$, the mentioned threshold field decreases at $T$ tending to $T_c$.

The theoretically described amplitude dependence of dielectric permittivity and also the tangent of the angle of dielectric losses tg$\delta$ were experimentally observed in many studies. Detailed research of these dependences were carried out, in particular, in [251,252] devoted to the study of amplitude dependence of dielectric losses in the crystal of triglycine sulphate doped with a chromium dopant and also exposed to $x$-rays. The results of these investigations (Fig.7.15) not only confirm the very fact of existence of the threshold field but also the previously made predictions about its temperature and concentration dependences.

## 7.9. AGEING AND DEGRADATION OF FERROELECTRIC MATERIALS

One of the most important processes, which restricts or at least hampers the practical application of ferroelectrics and related

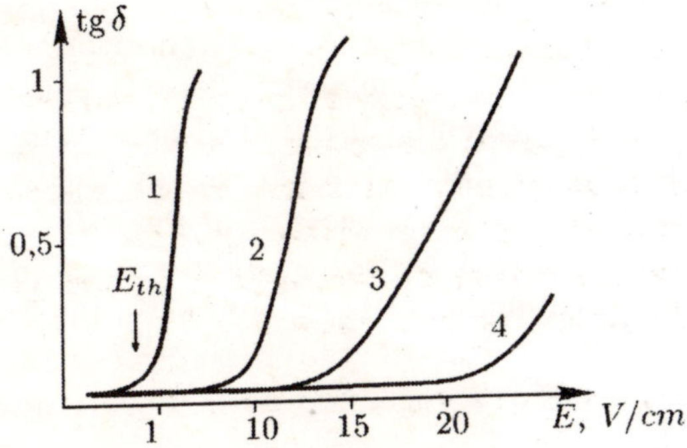


**Fig.7.15.** Amplitude dependence of tg$\delta$ at the frequency of 1 kHz and temperature of 46°C for a nominally pure crystal of TGS-1, for a crystal exposed to $x$-rays with the doses of 0.08 krad and 1.8 krad (curves 2 and 3, respectively), and for a crystal doped with chromium – 4.

materials is their ageing. The latter means the change (usually a decrease) of the characteristics of the material with time. From the practical viewpoint another important characteristic is the degradation of the materials used – a decrease of the service characteristics during operational use.

The nature of the above phenomena can be associated with many reasons. The degradation of the properties can be linked with mechanical processes, in particular, microcracking in the contact area of a ferroelectric with electrodes [253–257]. Both the degradation and ageing can be caused by diffusion processes in both the bulk of the material and on its surface.

When studying the process of ageing of a ferroelectric from the viewpoint of its dielectric properties, which, as shown above, are determined mainly by the displacements of domain boundaries, it is natural to consider primarily the interaction of the domain boundaries with the defects of the crystalline lattice, which controls these displacements. In a ferroelectric material with a freshly formed domain structure the lattice defects are distributed statistically uniformly throughout the volume of the material. In this case, the number of points of the boundary pinning by the defects is described by expression (5.14) and the domain contribution to dielectric permittivity by equation (6.6) correspondingly, which will be denoted hereinafter as

$$\varepsilon(t=0)=\frac{8P_0\sqrt{P_0}\tilde{\varepsilon}^{1/4}}{(\gamma a)^{1/4}\,n\mathcal{U}_0^{1/2}d}. \qquad (9.1)$$

During long term storage of a ferroelectric material in the polar phase it is natural to expect the redistribution of the defect locations in the volume of the material. This is caused, as shown above, mainly by the preferential location of defects in the boundary due to the decrease of the total energy of the system under consideration in this case. Thus, because of the diffusion of defects to the boundary we should expect an increase of the number of pinning points of the domain walls by defects in the course of time and decrease of the value of the dielectric constant of the ferroelectric material in accordance with the already mentioned expression (9.1).

When considering this phenomenon in our case, as in the description of the dielectric dispersion of the domain origin, it is natural to restrict ourselves to the same kinetic equation with the single relaxation time $\tau$:

$$\frac{dn}{dt}=-\frac{n-n_\infty}{\tau}, \tag{9.2}$$

in which the equilibrium concentration of defects $n_\infty(x)$ is no longer depends on time.

Taking into account this condition in our case solution (9.2) has a simple form

$$n(x,t)=n(x,t=0)\cdot e^{-t/\tau}+n_\infty(x)\left(1-e^{-t/\tau}\right). \tag{9.3}$$

Here $n(x,t=0)$ = $n_0$ and the dependence $n_\infty(x)$ taking into account the deformation of the profile of the domain wall as a result of its .capture by the neighbouring defects is

$$n_\infty(x)=n_0\exp\left(-\frac{\mathcal{U}(x)}{T}\right)=\begin{cases} n_0\exp\left(\dfrac{\mathcal{U}_0-\vartheta x^2/2}{T}\right), x<\tilde{U}, \\ n_0,\ x>\tilde{U}. \end{cases} \tag{9.4}$$

The condition determining the relation between $n_0$ and $l^2$ in our case is

$$\int_0^{\tilde{U}} n(x)dx=1/l^2. \tag{9.5}$$

Substituting here $n(x)$ in the form of (9.3), we obtain this relation in the form

$$n_0\tilde{U}\left[e^{-t/\tau}+\left(1-e^{-t/\tau}\right)\sqrt{\frac{\pi}{2}}\sqrt{\frac{T}{\mathcal{U}_0}}\cdot e^{\mathcal{U}_0/T}\times \right.$$
$$\left.\times\operatorname{erf}\left(\mathcal{U}_0/T\right)\right]=1/l^2. \tag{9.6}$$

The obtained expression differs from the self-consistence condition $n_0\widetilde{U}=1/l^2$, used when determining the initial value of the dielectric permittivity $\varepsilon(t=0)$ by the presence of the time depending expression in square brackets. Taking this into account and substituting into equation (9.1) the entire left part of equation (9.6) instead on $n_0$, for the time dependence of dielectric permittivity in the 'ageing' ferroelectric we obtain the following:

$$\varepsilon(t)=\frac{\varepsilon(t=0)}{\left[e^{-t/\tau}+\left(1-e^{-t/\tau}\right)\sqrt{\pi/2}\sqrt{T/\mathcal{U}_0}\cdot e^{\mathcal{U}_0/T}\operatorname{erf}\left(\mathcal{U}_0/T\right)\right]}. \tag{9.7}$$

As shown by equation (9.7), the ageing effect itself here depends only on the mobility of defects and energy of their interaction with domain walls.

According to considerations of the previous section, another important characteristics, which determines the dielectric properties of ferroelectrics, is the field of detachment of domain boundaries from defects. This field is of the threshold nature for the beginning of the amplitude dependence of the dielectric properties and in the case when the stage of detachment of domain boundaries from defects controls the switching processes, it can be regarded as a coercive field. The time dependence of a number of pinning points of the boundary by defects, described by equation (9.6) will evidently lead to a change in the detachment field of the domain boundaries from the defects as well. To determine the time dependent pressure on the boundary $F[t]$ from the direction of the defects let us substitute into the already known expression

$$F[U]=\int\frac{\partial\mathcal{U}(x-U)}{\partial x}n(x,t)\,dx=\int\frac{\partial\mathcal{U}(x)}{\partial x}n(x+U,t)\,dx \tag{9.8}$$

the obtained expression (9.3). This yields

$$F[t]=\frac{n_0\exp(-t/\tau)\vartheta}{2}\left[2\tilde{U}U-U^2\right]+$$

$$+2n_0\left(1-\exp(-t/\tau)\right)T\exp(-\vartheta U^2/2T)\operatorname{sh}\left(\frac{\vartheta\tilde{U}U}{T}\right). \tag{9.9}$$

Because of the cumbersome nature of the general equation (9.9), it is convenient to carry out the determination of the threshold (coercive) field in the analytical form here in two limiting cases: for the completely unaged ($t = 0$) and completely aged ($t\to\infty$) specimen. In the first case, this expression has already been determined and is represented by equation (9.2), in the second case on the basis of condition $2P_0E_c=F_{max}$ and the maximum force of interaction of the boundary with the defects, determined from (9.9) for ($t\rightarrow\infty$) it turns out to be as follows

$$E_c = \frac{n_0 T}{P_0} \cdot e^{\mathcal{U}_0/T}. \tag{9.10}$$

As expected, here the field $E_c$ increases with time and its specific strength depends on the ratio between $\mathcal{U}_0$ and $T$.

In the process of operation of the ferroelectric materials in low-amplitude external fields, like in the ageing process, the boundary is enriched by defects as a result of additional capture of the defects by the boundary from the region within which the boundary moves in the given external field. This also results in the increase of the number of pinning points of the boundary by defects, and, consequently, in the decrease of the contribution of domain boundaries to the magnitude of dielectric constant $\varepsilon$. At that the nature of distribution of the defects within the limits of the region, in which the domain boundary moves and, therefore, the strength of the resultant effect, depend on the amplitude and frequency of the applied fields.

## References


[1] Kanzig W. Ferroelectrics and Antiferroelectrics. New York; Acad. Press, 1957. - 234 p.

[2] Jona F., Shirane G., Ferroelectric crystals, New York: Macmillian, 1962. - 555 p.

[3] Zheludev I.S. Physics of crystalline dielectrics [in Russian]. Moscow: Nauka, 1968. - 464 p.

[4] Sonin A.S., Strukov B.A. Introduction to Ferroelectricity [in Russian]. Moscow: Vishchaya shkola, 1970. - 271 p.

[5] Kholodenko L.P. Thermodynamic Theory of $BaTiO_3$ - Type Ferroelectrics [in Russian]. Riga: Zinatne, 1972. - 227 p.

[6] Vaks V.G. Introduction to the Microscopic Theory of Ferroelectricity [in Russian]. Moscow: Nauka, 1973. - 327 p.

[7] Zheludev I.S. Foundations of ferroelectricity [in Russian]. Moscow: Atomizdat, 1973. - 472 p.

[8] Bursian E.V. Nonlinear crystal. Barium Titanate [in Russian]. Moscow: Nauka, 1974. - 296 p.

[9] Blinc R., Zeks B. Soft modes in ferroelectrics and anti-ferroelectrics. Amsterdam: North-Holland Publishing company, 1974.

[10] Fridkin V.M. Ferroelectrics Semiconductors, New York: Consultants Bureau, 1980.

[11] Strukov B.A. Ferroelectricity [in Russian]. Moscow: Nauka, 1979. - 96 p.

[12] Lines M.E. and Glass A.M. Principles and Application of Ferroelectrics and Related materials, - Oxford: Clarendon Press, 1977.

[13] Burfoot J., Taylor J. Polar dielectrics and their applications [in Russian]. Moscow: Mir, 1981. - 526 p.

[14] Izyumov Yu.A., Syromyatnikov V.N. Phase transitions and crystal symmetry [in Russian]. Moscow: Nauka, 1984. - 248 p.

[15] Bruce A.D., Cowley R.A. Structural phase transitions [in Russian]. Moscow: Mir. 1984. - 408 p.

[16] Smolenskii G.A., Bokov V.A., Isupov V.A. et al. Physics of ferroelectric phenomena [in Russian]. Leningrad: Nauka, 1985. - 396 p.

[17] Rudyak V.M. Switching processes in nonlinear crystals [in Russian]. Moscow: Nauka, 1986. - 243 p.

[18] Hubert A.. Theory of domain walls in ordered mediums [in Russian]. Moscow: Mir. 1987. - 306 p.

[19] Levanyuk A.P., Sigov A.S. Defects and structural phase transition. New York: Gordon and breach science publishers, 1988. - 208 p.

[20] Fesenko E.G., Gavrilyachenko V.G., Semenchev A.F. Domain structure of pluriaxial ferroelectric crystals [in Russian]. Roston/Don: Rostov State University, 1990. - 192 p.

[21] Strukov B.A., Levanyuk A.P. Physical Principles of Ferroelectric Phenomena in Crystals [in Russian]. Moscow: Nauka, 1995. - 301 p.

[22] Landau L.D., Lifshitz E.M. Electrodynamics of continuums [in Russian]. Moscow: Nauka, 1982. - 620 p.

[23] Klassen-Neklyudova M.V. Mechanical twinning of crystals [in Russian].

Moscow: Izd. AN SSSR, 1960. - 261 p.

[24] Garber R.I. Mechanism of twinning in tiff and alkaline saltpeter under plastic deformation. - Zh. Eksp. Teor. Fiz., 1947, V.17, N 1, P.48-62.

[25] Privorotskii M.A. On the theory of domain structures. - Zh. Eksp. Teor. Fiz., 1969, V.56, N 6, P.2129-2142.

[26] Privorotskii M.A. Thermodynamic theory of ferromagnetic domains. - Usp. Fiz. Nauk, 1972, V.108, N 1, P.43-80.

[27] Marchenko V.I. On ferroelectric domain structure. - Zh. Eksp. Teor. Fiz., 1979, V.77, N 61(12), P.2419-2421.

[28] Ingle S.G., Joshi S.C. Unstable point domains in ferroelectrics. - Phys. Rev. B, 1986, V.34, N 7, P. 4840-4845.

[29] Mitsui T., Furuchi J. Domain structure of rochelle salt and $KH_2PO_4$. - Phys.Rev., 1953, V.90, N 2, P.193-202.

[30] Fousek J., Safrankowa M. On the equilibrium domain structure of $BaTiO_3$. - Jap.J.Appl.Phys., 1965, V.4, N 6, P.404-408.

[31] Bulaevskii L.N. and Panyukov S.V. On the domain structure in the systems with non-local interaction. - Fiz Tverd. Tela (Leningrad), 1986, V.28, N 1, P.193-196.

[32] Roitburd A.L. Theory of formation of heterophase structure at phase transitions in solid state. - Usp. Fiz. Nauk, 1974, V.113, N 1, P. 69-104.

[33] Roitburd A.L. Influence of the mechanical stresses on the formation of domain structure during martensite and ferroelastic phase transitions. - Izv. Akad. Nauk SSSR, Ser.Fiz., 1983, V.47, N 3, P.435-448.

[34] Bar'yakhtar E.G., Bogdanov A.N., Yablonskii D.A. Physics of magnetic domains. - Usp. Fiz. Nauk, 1988, V.156, N 1, P.47-91.

[35] Sidorkin A.S., Sigov A.S. Unipolar state of ferroelectric plate with oriented polar defects. - Kristallografiya, 1999, V.44, N 1, P.109-110.

[36] Chenskii E.V. Termodynamic ratios for domain structure of ferroelectrics. - Fiz Tverd. Tela (Leningrad), 1972, V.14, N 8, P.2241-2246.

[37] Chenskii E.V., Tarasenko V.V. Theory of phase transitions in inhomogeneous state in the limited ferroelectrics in external electric field. – Zh. Eksp. Teor. Fiz., 1982, V.83, N 3(9), P.1089-1099 [Sov. Phys. JETP, 1982, V.56, N 3(9), P.618-628].

[38] Andreev A.F. Striction superstructure in two-dimensional phase transitions. – Pi'sma Zh. Eksp. Teor. Fiz., 1980, V.32, N 11, P. 654-656.

[39] Dikshtein I.E. Surface waves and domain structure in the ferroelastic film fixed on the deformable substrate in the vicinity of phase transition. - Fiz Tverd. Tela (Leningrad), 1986, V.28, N 6, P.1748-1752.

[40] Syrkin E.S., Gelfgat I.M. Deceleration of surface acoustic waves in the system film-substrate in the vicinity of structural phase transition. – Poverkhnost, 1982, V.1, N 8, P.28-29.

[41] Dikshtein I.E., Tarasenko V.V. Origination of domain structure in ferroelastic film on the massive substrate in the vicinity of phase transition. - Fiz Tverd. Tela (Leningrad), 1988, V.30, N 1, P.200-204.

[42] Darinskii B.M., Lazarev A.P., Sidorkin A.S. Formation of the domain structure by the cooling of ferroelectric surface. - Ferroelectrics, 1989, V. 98, P.193-195.

[43] Darinskii B.M., Lazarev A.P., Sidorkin A.S. Formation of the domain structure in ferroelastic sample with free surface. - Izv. Akad. Nauk SSSR, Ser.Fiz., 1989, V.53, N 7, P.1276-1279.

[44] Darinskii B.M., Lazarev A.P., Sidorkin A.S. Formation of the domain structure

in ferroelectric films with free charge carriers. - Kristallografiya, 1991, V.36, N.3, P.757-758 [Sov. Phys. Crysllogr., 1991, V.36, P.423].

[45] Sidorkin A.S., Darinskii B.M., Sigov A.S. Evolution of domain structure in ferroelectrics in the presence of polarization screening by charges bound to surface states and by free charge carriers. - Fiz Tverd. Tela (St. Peterburg), 1997, V.39, N 5, P.922-924 [Phys.Solid State, 1997, V.39, N 5, P.823-825].

[46] Tarasenko V.V., Chenskii E.V., Dikshtein I.E. Theory of flat domain structure in rare-earth orthoferrite in the vicinity of spin re-orientation point. - Fiz Tverd. Tela (Leningrad), 1976, V.18, N 6, P.1576-1582.

[47] Dikshtein I.E., Lisovskii F.V., Mansvetova E.G., Tarasenko V.V. Types of instabilities in ordered domain structures. - Zh. Eksp. Teor. Fiz., 1991, V.100, N 1(7), P. 205-223.

[48] Darinskii B.M., Lazarev A.P., Sidorkin A.S. Domain structure in an obliquely cut ferroelectric plate. – Fiz Tverd. Tela (Leningrad), 1993, V.35, N 7, P.1942-1946 [Phys. Solid State, 1993, V.35, N.7, P.969-971].

[49] Shur V.Ya., Gruverman A.L., Letuchev V.V., Rumyantsev E.L., Subbotin A.L. Domain structure of lead germanate. - Ferroelectrics, 1989, V.98, P.31-50.

[50] Darinskii B.M., Krivonos A.A., Lazarev A.P., Savvinov A.M., Sidorkin A.S. Formation of the domain structure in the Rochelle salt crystals $(NaKC_4H_4O_6)$ 4 $H_2O$ under irregular cooling. - Izv. Akad. Nauk SSSR, Ser.Fiz., 1993, V.57, N 9, P.63-65.

[51] Khachaturyan A.G. Theory of phase transformations and solid solution structure [in Russian]. - Moscow: Nauka. 1974. - 384 p.

[52] Zhirnov V.A. On the theory of the domain walls in ferroelectrics. – Zh. Eksp. Teor. Fiz., 1958, V.35, N 5, P.1175-1180 [Sov. Phys. JETP, 1958, V.8, N 5, P.822-827].

[53] Ivanchik I.I. On the microscopic theory of ferroelectrics. - Fiz Tverd. Tela (Leningrad), 1961, V.3, N 12, P.3731-3742.

[54] Bulaevskii L.N. Thermodynamic theory of domain walls in ferroelectrics of perofskite type. - Fiz Tverd. Tela (Leningrad), 1963, V.5, N 11, P.3183-3187.

[55] Darinskii B.M., Nechaev V.N. Electric field in the 90-degree domain boundaries of ferroelectrics with perofskite structure. - Fiz Tverd. Tela (Leningrad), 1979, V.21, N 2, P.520-523.

[56] Darinskii B.M., Lazarev A.P., Sidorkin A.S. Structure of a domain boundary in the vicinity of the surface of a ferroelectric. – Fiz Tverd. Tela (Leningrad), 1989, V.31, N 11, P.287-289 [Sov. Phys. Solid State, 1989, V.31, N 11, P.2003].

[57] Darinskii B.M., Fedosov V.N. Structure of an interphase boundary in perofskite type ferroelectrics. - Fiz Tverd. Tela (Leningrad), 1974, V.16, P.594-595.

[58] Dmitriev S.G. On the problem of screening of spontaneous polarization field by free charge carriers in ferroelectrics. – Zh. Eksp. Teor. Fiz., 1980, V.78, N 1, P.412-419.

[59] Darinskii B.M., Fedosov V.N. Structure of interphase boundary in ferroelectrics semiconductors. – Fiz. Tekh. Poluprovodnikov, 1978, V.12, N 3, P.603-605.

[60] Meleshina V.A., Indenbom V.L., Bagdasarov H.S., and Polkhovskaya T.M. Domain walls, antiphase boundaries and dislocations in Gadolinium Molybdate Crystals. - Kristallografiya, 1973, V.18, N 6, P.1218-1226.

[61] Ishibashi Y., Dvorak V. Domain walls in Improper Ferroelectrics. - J. Phys.Soc.Jap., 1976, V.41, N 5, P.1650-1658.

[62] Darinskii B.M., Nechaev V.N. Interaction of domain and antiphase boundaries in improper ferroelectrics. - Izv. Akad. Nauk SSSR, Ser.Fiz., 1979, V.

[63] Ishibashi Y. Phenomenological theory of domain walls. - Ferroelectrics, 1989, V.98, P.193-205.

[64] Zaleskii A.M., Savvinov A.M., Zheludev I.S., Ivanchenko A.I. Nuclear magnetic resonance in domain boundaries of $DyFeO_3$ Crystal. - Fiz Tverd. Tela (Leningrad), 1973, V.15, P.93-98.

[65] Bogdanov A.N., Galushko V.A., Telepa V.T., Yablonskii D.A. Spin re-orientation in 180-degree domain boundaries of spin-flop phase of easy – axis ferromagnetics. – Pi'sma Zh. Eksp. Teor. Fiz., 1984, V.40, N 11, P.453-455.

[66] Bogdanov A.N., Telepa V.T., Shatskii P.P., Yablonskii D.A. Phase transitions induced by external field in the domain walls of rhombic antiferromagnetic $(C_2H_5NH_3)_2$. – Pi'sma Zh. Eksp. Teor. Fiz., 1986, V.90, P.1738-1747.

[67] Khlyustikov I.N., Khaikin M.S. Twinning interface of metal crystal – two-dimentional superconductor. – Pi'sma Zh. Eksp. Teor. Fiz., 1981, V.33, P.167-170.

[68] Khlyustikov I.N., Khaikin M.S. Superconductivity of the layer of microscopic twin-crystals. - Pi'sma Zh. Eksp. Teor. Fiz., 1981, V.34, N 4, P.207-211.

[69] Bullbich A.A., Gufan Yu.M. Phase transition in domain walls. - Ferroelectrics, 1989, V.98, P.277-290.

[70] Sonin E.B., Tagantsev A.K. Structure and phase transition in antiphase boundaries of improper ferroelectrics. - Ferroelectrics, 1989, V.98, P.291-295.

.[71] Kahn J. Theory of the crystal growth and movement of interphase boundary in crystalline materials. – Usp. Fiz. Nauk, 1967, V.91, P.677-689.

[72] Darinskii B.M., Fedosov V.N. The motion of 180 – degree domain boundary in ferroelectrics of perofskite type. - Izv. Akad. Nauk SSSR, Ser.Fiz., 1971, V.35, N 9, P.1795-1797.

[73] Seeger A. And Schiller P. Kinks on dislocations and their influence on the internal friction in crystals. - in: Physical Acoustics. Principles and Methods (ed. Mason W.P.), V.3, Part A, Academic Press, New York, 1966. – 361 p.

[74] Lawless W.N. 180° domain wall in $BaTiO_3$. - Phys. Rev., 1968, V.175, N 2, P.619-624.

[75] Strokach A.A. Microscopic consideration of 180-degree domain walls in perofskite. – Ukr.Fiz.Zh., 1981, V.26, N 4, P.656-659.

[76] Kinase W., Orayama H., Yoshikawa A., Taguchi N. On the 90° domain wall of $BaTiO_3$ crystal. - J. Phys. Soc. Japan, 1970, V.28, Suppl., P.383-385.

[77] Darinskii B.M., Fedosov V.N. The structure of 90° domain boundary in $BaTiO_3$. - Fiz Tverd. Tela (Leningrad), 1971, V.13, N 1, P.22-27.

[78] Bjorkstam J.L., Oettel R.E. The effect of deuteron substitution on domain dynamics in potassium dihydrogen phosphate (KDP). - Proc. Intern. Meet. Ferroelectricity, Prague, 1966, V.2, P.91-96.

[79] Fedosov V.N., Sidorkin A.S. Influence of tunneling on the structure of domain boundaries in $KH_2PO_4$ crystal. - Izv. Akad. Nauk SSSR, Ser.Fiz., 1975, V.39, N 4, P.682-685.

[80] Tokunaga M., Matsubara T. Theory of Ferroelectric Phase Transition in

$KH_2PO_4$ Type Crystals. I. - Progress of Theoretical Physics, 1966, V.35, N 4, P.581-599.

[81] Novakovic L. A possible theoretical explanation of the shift of the Curie temperature in $KH_2PO_4$ and $KD_2PO_4$ upon a hydrostatic pressure. - J. Phys. Chem. Sol, 1968, V.29, P.963-966.

[82] Moore M.A., Willians H.C.W.L. Theory of hydrogen-bonded ferroelectrics: I. - J.Phys. C: Solid State Phys., 1972, V.5, P.3168-3184.

[83] Samara G.A. The effects of deuteration on the static ferroelectric properties of $KH_2PO_4$ (KDP). - Ferroelectrics, 1973, V.5, P.25-37.

[84] Strukov B.A., Vaks V.G. Baddur A., Zinenko V.I., Koptsik V.A. Experimental and theoretical study of phase transitions in $KH_2PO_4$ (KDP)-type crystals. - Ferroelectrics, 1974, V.7, P.195-197.

[85] Lawrence M.C., Robertson G.N. Estimating the barrier to proton or deuteron tunnelling in hydrogen-bonded crystals. - Ferroelectrics, 1980, V.25, N 1-4, P.363-366.

[86] Kamysheva L.N., Fedosov V.N., Sidorkin A.S. Temperature Dependence of the Domain Wall Mobility in the $KH_2PO_4$ (KDP) Group Crystals. - Ferroelectrics, 1976, V.13, N 1-4, P.463.

[87] Fedosov V.N., Sidorkin A.S. Effect of two-dimensional ordering on the mobility of domain boundaries. - Fiz Tverd. Tela (Leningrad), 1977, V.19, N 8, P.1322-1326 [Sov.Phys. Solid State, V.19, N 8, P.1359-1361.

[88] Tentrup T., Weyrich K.H., Siems R. Domain walls in pseudo Spin Systems with Competing interactions. - Japanese Journal of Applied Physics, 1985, V.24, S.24-2, P.571-573.

[89] Sidorkin A.S. Influence of tunneling on the structure and mobility of domain walls in order-disorder ferroelectrics. – Fiz Tverd. Tela (Leningrad), 1989, V.31, N 9, P.293-295 [Sov. Phys. Solid State, 1989, V.31, N 9, P.1645-1646].

[90] Egami T., Theory of Bloch Wall Tunneling. - Phys. Stat. Sol., 1973, V.57, N 1, P.211-224.

[91] Baldwin J.A., Milstein F., Wong R.C. Slow magnetic domain wall motion and tunneling. - J. Appl. Phys., 1977, V.48, N 1, P.396-397.

[92] Vaks V.G., Zein N.E. Ferroelectric properties of solid solutions $KH_{2x}D_{2(1-x)}PO_4$-type. - Fiz Tverd. Tela (Leningrad), 1975, V.17, N 6, P.1617-1626.

[93] Holakovsky J. Study of phase transition in mixed $KH_{2x}D_{2(1-x)}PO_4$ crystals by modified molecular field approximation. - Czech. J. Phys. - 1972, V. 22, N 8, P.651-665.

[94] Sidorkin A.S., Fedosov V.N., Burdanina N.A., Kamysheva L.N. Influence of the domain structure on switching processes in a $KH_{2x}D_{2(1-x)}PO_4$ (DKDP) crystal. - Fiz Tverd. Tela (Leningrad), 1979, V.21, N 3, P. 861-865 [Sov. Phys. Solid State, 1979, V.21, N 3, P.504- 507].

[95] Popov E.S., Shuvalov L.A. Surface tension of 180-degree domain walls in some phenomena in triglycine sulphate and c-domain barium titanate crystals. - Kristallografiya, 1973, V.18, N 3, P.642-644.

[96] Sidorkin A.S., Fedosov V.N. Ionization of charged impurity centers due to polarization reversal in a ferroelectric. - Fiz Tverd. Tela (Leningrad), 1981, V.23, N 9, P.2854-2856 [Sov.Phys. Solid State, 1981, V.23, N 9, P.1667- 1668].

[97] Sidorkin A.S., Darinskii B.M. The structure of domain boundary in ferroelectrics with point defects. - Fiz Tverd. Tela (Leningrad), 1986, V.28,

N 1, P.285-288.

[98] Abramowitz and Stegun I.A. (eds.), Handbook on Mathematical Functions with Formulas, Graphs, and Mathematical Tables, New York: Dever, 1964.

[99] Landau L.D. and Lifshitz. Theory of Elasticity [in Russian]. Moscow: Nauka, 1965. - 202 p.

[100] Hirth J.P., Lothe J. Theory of dislocations[in Russian]. Moscow: Atomizdat, 1972. – 599p.

[101] Kosevich A.M. Dislocations in the Theory of elasticity [in Russian]. Kiev: Naukova Dumka, 1978. - 219 p.

[102] Eshelby J.D. Continual theory of dislocations [in Russian]. Moscow: IL, 1963. - 247 p.

[103] De Wit R. Continual theory of disclinations [in Russian]. Moscow: Mir. 1977. - 208 p.

[104] Kosevich A.M., Boiko V.S. Dislocation theory of elastic twinning of the crystals. - Usp. Fiz. Nauk, 1971, V.104, N 2, P.201 - 254.

[105] Boiko V.S., Kosevich A.M., Feldman E.P. Dislocation description of the wall of elastic domains. - Fiz Tverd. Tela (Leningrad), 1987, V.29, N 1, P. 170-177.

[106] Pertsev N.A., Arlt G. Dislocation method of calculation of the internal stresses in polycrystalline ferroelastics. - Fiz Tverd. Tela (Leningrad), 1991, V.33, N 10, P.3077-3088.

[107] Sidorkin A.S., Shevchenko A.A. The interaction of domain wall in ferroelectric crystal with dilatation center. - Ferroelectrics, 1993, V.141, P.235-242.

[108] Kosilov A.T., Perevoznikov A.M., and Roshchupkin A.M. Interaction of coherent interphase boundaries with point defects in crystals. - Poverkhnost, 1984, V.58, N 1, P.5-10.

[109] Neel L. Influence of the cavities and inclusions on the coercive force. – in: Physics of ferromagnetic regions. Moscow: IL, 1951. P.215-239.

[110] Prause W. The interaction energy between a spherical cality and ferromagnetic domain wall. - J. Magn. and Magn. Mat., 1977, V.4, P.344-355.

[111] Lazarev A.P., Fedosov V.N. Depolarization field of finite size inclusion in ferroelectric. - Kristallografiya, 1984, V.29, N 6, P.1182-1184.

[112] Kosilov A.T., Perevoznikov A.M., Poshchupkin A.M. Interaction of coherent interphase boundaries with dislocations in crystals. - Poverkhnost, 1983, N 9, P.25-30.

[113] Nechaev V.N. Interaction of dislocation with a domain boundary in a ferroelectric. -Fiz Tverd. Tela (Leningrad), 1991, V.33, N 5, P.1536-1568.

[114] Gorbunov V.V., Darinskii B.M. Bending of a domain boundary in the field of inclined dislocation in ferroelastic. - Izv. Akad. Nauk SSSR, Ser.Fiz., 1989, P.53, N 7, P.1292-1295.

[115] Zheludev I.S., Shuvalov L.A. Ferroelectric phase transitions and crystal symmetry. - Kristallografiya, 1956, V.1, N 4, P. 681-688.

[116] Zheludev I.S., Shuvalov L.A. Domain orientation and macro-symmetry of the properties of ferroelectric crystals. - Izv. Akad. Nauk SSSR, Ser.Fiz., 1957, V.21, N 2, P.264-274.

[117] Fousek J., Janovec V. The orientation of domain walls in twinned ferroelastic crystals. - J.Appl.Phys., 1969, V.40, N 1, P. 135-142.

[118] Fousek J. Permissible domain walls in ferroelectric species. - Chech. J.Phys., 1971, V.B21, N 9, P.955-968.

[119] Sapriel J. Domain-wall orientations in ferroelastics. - Phys. Rev. B., 1975, V.12, N 11, P.5128-5140.

[120] Hatano J., Suda F., Futama H. Orientation of the Ferroelectric domain wall in Triglycine Sulphate Crystals. - J. Phys. Soc. Japan, 1976, V.41, N 1, P.188-193.

[121] Shuvalov L.A., Dudnik E.F., Wagin S.V. Domain structure geometry of real ferroelastics. - Ferroelectrics, 1985, V.65, N 1-2, P.143-152.

[122] Parker T.J., Barfoot I.C. The structure of transition fronts in barium titanate. - Brit. J. Appl.Phys., 1966, V.17, N 2, P.213-218.

[123] Barfoot I.C., Parker T.J. The movement of transition fronts in barium titanate. - Brit. J. Appl.Phys., 1966, V.17, N 2, P.219-222.

[124] Fedosov V.N., Sidorkin A.S. "Fine" structure of a phase boundary in a ferroelectric. - Fiz Tverd. Tela (Leningrad), 1979, V.21, N 7, P.2116-2120 [Sov. Phys. Solid State, 1979, V.21, N 7, P.1211-1214].

[125] Krainyuk G.G., Otko A.I. Statics of a zig-zag boundary in gadolinium molybdate. - Kristallografiya, 1989, V.34, N 2, P.502-504.

[126] Natterman T. Metastability and Depinning Threshold in the Random Field Ising Model. - Phys. Stat.Sol.(b), 1985, V.129, P.153-162.

[127] Natterman T. Dielectric Tails in $K_2ZnCl_4$: No Evidence for an Intrinsic Chaotic State. - Phys. Stat.Sol.(b), 1986, V.133, P.65-69.

[128] Rice T.M., Whitchaise S.W., Littleword P. Impurity pinning of discommensuraties in charge-density waves. - Phys. Rev.B, 1981, N 5, P.2751-2759.

[129] Roitburd A.L. Instability of regions in the vicinity of the boundary and formation of zig-zag interdomain and interpfase boundaries. - Pi'sma Zh. Eksp. Teor. Fiz., 1988, V.74, N 3, P.141-143.

[130] Chervonobrodov S.P., Roitburd A.L. Orientation instability of domain boundary in ferroelectrics. - Ferroelectrics, 1988, V.93, N 1-4, P.109-112.

[131] Darinskii B.M., Gorbunov V.V. Instability of inclined 180° domain boundary in ferroelectrics. - Ferroelectrics, 1989, V.89, P.235-240.

[132] Darinskii B.M., Sidorkin A.S., Lazarev A.P. The structure and the motion of the domain baundaries of ferroelectrics-ferroelastics. - Ferroelectrics, 1989, V.98, P.245-252.

[133] Bastie P. Bornarel J., Legrand J.F. Pertubated region in the vicinity of domain wall in ferroelectric - ferroelectric crystals. - Ferroelectrics, 1976, V.13, N 1-4, P.455-458.

[134] Tylczynski Z. Determination of domain wall thickness in TGS crystals from measurements of longitudinal ultrasonic wave propagation. - Acta Phys. Polon., 1977, V.A51, N 6, P. 865-868.

[135] Papria P.Y. Determination of the thickness of the ferroelectric domain boundary in triglicine sulphate by X-ray topography. - Phil. Mag., 1982, V. A 46, N 4, P.691-697.

[136] Petroff J.F. Topographyc Study of 180° Domains in Triglicine Sulphate. - Phys. Stat. Sol (b), 1969, V.31, P.285-295.

[137] Yamamoto N. Yagi K., Honjo G. Electron microscopic observation of ferroelectric and antiphase domain in $Gd_2(MoO_4)_3$. - Phyl. Mag., 1974, 8 ser., V.30, N 5, P.1161-1164.

[138] Suzuki K. Observation of diffracted light from domain wall in $Gd_2(MoO_4)_3$ - Sol. Stat. Comm., 1972, V.11, P.937-939.

[139] Esayan S.H., Lemanov V.V., Smolenskii G.A. Reflection and refraction of elastic waves on domain boundaries in $Gd_2(MoO_4)_3$ ferroelectric crystal. – Dokl. AN SSSR, 1974, V.217, N 1, P.83-85.

[140] Lin Peng Ju, Bursill L.A. On the width of the charged {110} ferroelec-

tric domain walls in potassium niobate. - Phil. Mag., 1983, V.48, N 2, P.251-263.

[141] Darinskii B.M., Sidorkin A.S. Effective width of domain wall in ferroelectric crystals with defects. - Fiz Tverd. Tela (Leningrad), 1984, V.26, N 11, P.3410-3414 [Sov.Phys. Solid State, 1984, V.26, N 11, P.2048-2050].

[142] Sidorkin A.S., Darinskii B.M., Pachevskaya G.N. Effective width of domain wall in ferroelastics with defects. – Izv. Akad. Nauk SSSR, Ser.Fiz., 1987, V.51, P.389-392.

[143] Sannikov D.G. On the theory of domain boundaries motion in ferroelectrics. - Izv. Akad. Nauk SSSR, Ser.Fiz., 1964, V.28, N 4, P.703-707.

[143] Krumhansl J.A., Schrieffer J.R. Dynamics and statistical mechanics of a one-dimensional model Hamiltonian for structural phase transitions. - Phys.Rev.B., 1975, V.11, N 9, P.3535-3545.

[145] Bishop A.R., Domani E., Krumhansl J.A. Quntum correction to domain walls in a model (one-dimensional) ferroelectric. - Phys. Rev. B., 1976, V.14, N 7, P.2966-2971.

[146] Pouget J., Maugin G.A. Influence of an external electric field on the motion of a ferroelectric domain wall. - Physics Letters, 1985, V.109A, N 8, P.389-392.

[147] Pouget J., Maugin G.A. Solitons and electroacoustic interactions in ferroelectric crystals. I. Single solitons and domain walls. - Phys. Rev B, V.30, N 9, P.5306-5325.

[148] Pouget J., Maugin G.A. Solitons and electroacoustic interactions in ferroelectric crystals. II. Interactions of solitons and radiations. - Phys. Rev B, V.31, N 7, P.4633-4649.

[149] Darinskii B.M., Parshin A.V., Turkov S.K. Phonon mechanism of deceleration of 90-degree domain boundaries in ferroelectrics. - Izv. Akad. Nauk SSSR, Ser.Fiz., 1975, V.39, N 4, P.677-681.

[150] Drougard M.E. Detailed study of switching current in barium titanate. - J. Appl. Phys., 1960, V.31, N 2, P.352-355.

[151] Miller R.C., Weinreich G. Mechanism for the Sidewise Motion of 180-degree Domain Walls in Barium Titanate. - Phys. Rev., 1960, V.117, N 6, P.1460-1466.

[152] Flippen R.B. Domain Wall dynamic in ferroelectric/ferroelastic molybdates. - J. Appl. Phys., 1975, V.46, N 3, P.1068-1071.

[153] Hayashi M. Kinetics of domain wall motion in ferroelectric switching. I. General formulation. - J.Phys. Soc. Japan, 1972, V.33, N 3, P.616-628.

[154] Hayashi M. Kinetics of domain wall motion in ferroelectric switching. II. Application to Barium Jitanate. - J.Phys. Soc. Japan, 1973, V.34, N 5, P.1240-1244.

[155] Burtsev E.V., Chervonobrodov S.P. The possible model for description of the switching process of ferroelectrics in a weak electric field. - Kristallografiya, 1982, V.27, N 5, P.843-850.

[156] Darinskii B.M., Sidorkin A.S. The motion of the domain boundaries in crystals of $KH_2PO_4$ Group. Ferroelectrics, 1987, V.71, P.269-279.

[157] Landau L.D. Collected works [in Russian]. Moscow: Nauka, 1969, Part 2, 450 p.

[158] Tikhomorova N.A., Chumakova S.P., Shuvalov L.A. et al. Use of liquid crystals for studying of switching processes in ferroelectrics. Kristallografiya. 1987, V.32, N 1, P.143-148.

[159] Sogr A.A., Borodin V.Z. Observation of the domain structure dynamics

of ferroelectrics in a focused beam microscope. - Izv. Akad. Nauk SSSR, Ser.Fiz., 1984, V.48, N 6, P.1086-1089.

[160] Bulbich A.A., Dmitriev V.P., Zhelnova O.A., Pumpyan P.E. Nucleation on the twin boundaries in crystals, undergoing the structural phase transitions in the vicinity of critical points of phase diagrams. Direct observation of nucleation of low-symmetry phases in $NaNbO_3$. - Zh. Eksp. Teor. Fiz., 1990, V.98, N 3(9), P.1108-1119.

[161] Korzhenevskii A.L. On phase transitions of the first order in the dislocation crystals. - Fiz Tverd. Tela (Leningrad), 1986, V.28, N 5, P.1324-1331.

[162] Nabutovskii V.M., Shapiro B.Ya. Superconducting string in the vicinity of dislocation. - Zh. Eksp. Teor. Fiz., 1978, V.75, N 3, P.948-959.

[163] Bulbich A.A., Pumpyan P.E. Local phase transition on the moving dislocation.- Kristallografiya, 1989, V.35, N 2, P.284-288.

[164] Barkla H.M., Finlayson D.M. The properties of $KH_2PO_4$ bellow the Curie point. - Phyl. Mag., 1953, V.44, N 349, P.109-130.

[165] Bornarel J., Fouskova A., Gudon P. And Lajzerowicz J. Electrycal and optical investigation of ferroelectric properties of $KH_2PO_4$. - Proc. Intern. Meet Ferroelectricity, Prague, 1966, V.11, P.81-90.

[166] Kamysheva L.N., Zhukov O.K. On the non-linear properties $KH_2PO_4$. - Proc. Intern. Meet Ferroelectricity, Prague, 1966, V.11, P.200-203.

[167] Shuvalov L.A., Zheludev I.S., Mnatsakanyan A.V., Ludupov Ts.Zh., and Fiala I. Ferroelectric anomalies of dielectric and piezoelectric properties of $KH_2PO_4$ and $KD_2PO_4$ crystals. – Izv. Akad. Nauk SSSR Ser. Fiz., 1967, V.31, P.1919-1924.

[168] Malek Z., Shuvalov L.A., Fiala I., and Shtraiblova Ya. On anomalous behaviour of dielectric permittivity of RDP crystals at temperatures below the Curie point. – Kristallografiya, 1968, V.13, N 5, P.825-830 [Sov. Phys. Crystallogr., 1969, V.13, P.713].

[169] Kamysheva L.N., Zolototrubov Yu.S., Gridnev S.A., Sidorkin A.S., Rez I.S. Dielectric properties of rubidium dihydrogen arsenate in the vicinity of phase transition temperature. - in: Mechanisms of Relaxation Phenomena in Solids [in Russian], Kaunas, 1974, P.273-276.

[170] Burdanina N.A., Kamysheva L.N., Gridnev S.A., Zolototrubov Yu.S., and Zhurov V.I. Dielectric and Mechanical losses in $KH_2PO_4$ Crystals with specially introduced defects. - in: Mechanisms of Relaxation Phenomena in Solids [in Russian], Kaunas, 1974, P.239-244.

[171] Peshikov E.V., Shuvalov L.A. Amplitude dependence of dielectric parameters of rubidium dihydrogen phosphate. - Kristallografiya, 1981, V.26, N 4, P.855-858 [Sov.Phys. Crystallogr. 1981, V.26, P.484].

[172] Sidorkin A.S., Burdanina N.A., Lazarev A.P. Amplitude dependence of the domain structure "freezing" effect. – Fiz. Tverd. Tela (Leningrad), 1986, V.26, N 5, P.1419-1423 [Sov. Phys. Solid State, V.26, N 5, P.861-863].

[173] Baranov A.I., Shuvalov L.A., Rabkin V.S., Rashkovich L.N. Critical behaviour of dielectric permittivity in uniaxial ferroelectric $CsH_2PO_4$. - Kristallografiya, 1979, N 3, P.524-527.

[174] Levstik A., Blinc R., Kadaba P., Cizikov S., Levstik I., Filipic C. Dielectric properties of the 250 K phase transition in $BaMnF_4$ and $CsH_2PO_4$. - Ferroelectrics, 1976, V.14, P.703-706.

[175] Uesu Y., Kobayashi J. Crystal structure and ferroelectricity of cesii dihydrogen phosphate $CsH_2PO_4$. - Phys. Stat. Sol (a), 1976, V.34, N 2, P.475-482.

[176] Rashkovich L.N., Matveeva K.B. On cesium dihydrogen phosphate properties. - Kristallografiya, 1978, V.23, N 4, P.796-800.

[177] Kamysheva L.N., Drozhdin S.N., Sidorkin A.S. The "freezing" of the domain structure in the CDP crystals resulting from the structural reconstruction of the domain walls. - Ferroelectrics, 1981, V.31, P.37-40.

[178] Kamysheva L.N., Drozhdin S.N., Sidorkin A.S., Bukhman V.S. Peculiarities of polarization properties of cesium dihydrogen phosphate determined by the presence of the domain structure. - Kristallografiya, 1981, V.26, N 3, P.540-545.

[179] Kamysheva L.N., Burdanina N.A., Zhukov O.K., Bespamyatnova L.A. Influence of strong electric field on dielectric properties of $KD_2PO_4$ crystal. - Kristallografiya, 1969, V.14, N 1, P.162-163 [Sov. Phys. Crystallogr., 1969, V.14, P.141].

[180] Sidorkin A.S., Burdanina N.A., Kamysheva L.N., Fedosov V.N. Influence of the domain structure on the switching processes in a $KH_{2x}D_{2(1-x)}PO4$ (DKDP) crystal. - Fiz Tverd. Tela (Leningrad), 1979, V.21, N 3, P.861-864 [Sov.Phys. Solid State, 1979, V.21, N 3, P.504-507].

[181] Kryzhanovskaya N.A., Dovchenko G.V., Dudnik E.F., and Mishchenko A.V. Switching processes in $Rb(D_xH_{(1-x)})_2PO_4$ crystals. – Izv. Akad. Nauk SSSR Ser. Fiz., 1975, V.39, P.1031-1035.

[182] Bornarel J., Lajzerowicz J. Experimental evidence for the long-range interaction of domain in ferroelectric $KH_2PO_4$. - J.Appl. Phys., 1968, V.39, N 9, P.4339-4341.

[183] Bornarel J. Lajzerowicz J. Interdomain and domain - defect interactions in KDP. - Ferroelectrics, 1972, V.4, P.177-187.

[184] Bornarel J. Domains in $KH_2PO_4$. Ferroelectrics, 1987, V.71, P.255-268.

[185] Sidorkin A.S. The Mobility of Domain walls in Hydrogen-containing ferroelectric crystals. - Ferroelectrics, 1993, V.150, P.313-317.

[186] Choi Byoung-Koo, Kim Jong-Jean. Ammonium dihydrogen phosphate (ADP) impurity effects on phase transition and domain-wall freezing in potassium-ammonium dihydrogen phosphate. Phys. Rev. 1983, V. B 28, N 3, P.1623-1625.

[187] Andrew F.R., Cowley R.A. X-ray scattering from critical actuations and domain walls in KDP and DKDP. - J. Phys. (C). Solid St.Phys., 1986, V.19, N 4, P.615-635.

[188] Kamysheva L.N., Fedosov V.N., Sidorkin A.S. Pyroelectric Anomalies rezulting from domain structure reconstruction in ferroelectrics. - Ferroelectrics, 1984, V.34, P.27-30.

[189] Privorotskii M.A. Surface electromagnetic waves on the domain boundaries in ferroelectrics. - Physics Letters, 1977, V. A 60, P.361-362.

[190] Laikhtman B.D. Bending vibrations of domain walls and dielectric dispersion in ferroelectrics. - Fiz Tverd. Tela (Leningrad), 1973, V.15, N 1, P.93-102.

[191] Kessenikh G.G., Lyubimov V.N., Sannikov D.G. Surface elastically - polarized waves on the domain boundaries in ferroelectrics. - Kristallografiya, 1972, V.17, N 3, P.591-594.

[192] Fedosov V.N. Spectrum of the domain boundaries vibrations in piezoelectric and magnetoacoustic mediums. - Fiz Tverd. Tela (Leningrad), 1988, V.30, N 9, P.2572-2574.

[193] Nechaev V.N., and Roshchupkin A.M. On dynamics of the domain boundaries in ferroelectrics and ferromagnetics. - Fiz Tverd. Tela (Leningrad), 1988,

V.30, N 6, P.1908-1910.

[194] Darinskii B.M., Sidorkin A.S., Kostsov A.M. Oscillations of the domain boundaries in ferroelectrics and ferroelastics with point defects. - Izv. Akad. Nauk SSSR, Ser.Fiz., 1991, V.55, N 3, P.583-590.

[195] Darinskii B.M., Sidorkin A.S., Kostsov A.M. Oscillatuions of the domain boundaries in real ferroelectrics and ferroelectrics-ferroelastics. - Ferroelectrics, 1994, V.160, P.35-45.

[196] Novatskii V. Electromagnetic effects in solids. Moscow: Mir, 1986. – 159p.

[197] Darinskii B.M., Sidorkin A.S. Oscillations of domain boundaries in ferroelectrics and ferroelastics. - Fiz Tverd. Tela (Leningrad), 1987, V.29, N 1, P.3-7; 1989, V.31, N 7, P.307.

[198] Darinskii B.M., Sidorkin A.S. Vibrations of domain walls in real ferroelastics. – Fiz Tverd. Tela (Leningrad), 1989, V.31, N 12, P.28-31 [Sov. Phys. Solid State, 1989, V.31, N 12, P.2041-2043].

[199] Kosevich A.M., Polyakov M.L. Collective vibrations of dislocation wall. - Fiz Tverd. Tela (Leningrad), 1979, V.21, N 10, P.2941-2946.

[200] Nechaev V.N. On bending vibrations of 90$^{\circ}$ boundaries in ferroelectrics. - Fiz. Tverd. Tela (Leningrad), 1990, V.32, N 7, P.2090-2093.

[201] Fousek J. The contribution of domain walls to the small-signal complex permittivity of $BaTiO_3$. - Czech.J.Phys., 1965, V. B15, N 6, P.412-417.

[202] Fousek J.Brzhesina B. Frequency dependencies of the 90-degree domain walls motion in Barium Titanate. - Izv. Akad. Nauk SSSR, Ser.Fiz., 1964, V.28, N 4, P.717-721.

[203] Fedosov V.N., Sidorkin A.S. Quasielastic displacements of domain boundaries in ferroelectrics. - Fiz Tverd. Tela (Leningrad), 1976, V.18, N 6, P.1661-1668.

[204] Malosemov A., Slonzuski J. Domain boundaries in the materials with cylindrical magnetic domains [in Russian]. Moscow: Mir, 1982. - 382 p.

[205] Gurevich L.E., Liverts E.V. Quasi-acoustical and quasi-optical oscillations of the domain structure of a uniaxial ferromagnetic.- Zh. Eksp. Teor. Fiz., 1982, V.82, N 1, P.220-223.

[206] Sidorkin A.S. Translation oscillations of domain structure in ferroelectrics and ferroelastics. - Ferroelectrics, 1996, V.186, N1-4, P.239-242.

[207] Sidorkin A.S. Dynamics of Domain Walls in ferroelectrics and ferroelastics. - Ferroelectrics, 1997, V.191, P.109-128.

[208] Sidorkin A.S. Translational vibrations of domain srtucture in ferroelectrics. - Journal of Applied Physics, 1998, V.83, N 7, P.3762-3768.

[209] Sidorkin A.S. Nesterenko L.P. Effective mass and intrinsic frequency of oscillations for the translational motion of 180$^{0}$ domain walls in ferroelectric and ferroelastic materials. – Fiz. Tverd. Tela (St. Peterburg), 1995, V.37, N 12, P.3747-3750 [Phys.Solid State, 1995, V.37, N.12, P.2067-2068].

[210] Roitburd A.L. Phase equilibrium in solid. - Fiz Tverd. Tela (Leningrad), 1986, V.28, N 10, P. 3051-3054.

[211] Arlt G., Pertsev N.A. Force constant and effective mass of 90$^{\circ}$ domain walls in ferroelectric ceramics. - J.Appl.Phys., 1991, V.70, N 4, P.2283-2289.

[212] Pertsev N.A., Arlt G. Forced translated vibrations of 90$^{\circ}$ domain walls and the dielectric dispersion in ferroelectric ceramics. - J. Appl. Phys. 1993, V.74, N 6, P. 4105-4112.

[213] Sidorkin A.S., Sigov A.S. Translational vibrations of domain boundaries in ferroelectrics with different defect concentration. - Ferroelectrics, 1998,

V.219, P.199-205.

[214] Fedosov V.N., Sidorkin A.S. Nonlinearity of dielectric permittivity in ferroelectric crystals with defects. - Fiz Tverd. Tela (Leningrad), 1977, V.19, N 6, P.1632-1637 [Sov. Phys. Solid State, 1977, V.19, P.952].

[215] Fedosov V.N., Sidorkin A.S. Mobility of domain boundaries in $KH_2PO_4$ (KDP) group crystals. - Kristallografiya, 1976, V.21, N.6, P.1113-1116 [ Sov. Phys. Crystallogr., 1976, V.21, P.644].

[216] Sannikov D.G. Dispersion of dielectric permittivity in ferroelectrics. - Zh. Eksp. Teor. Fiz., 1961, V.41, N 1, P.133-138.

[217] Nettleton R.E. Switching Resonance in Crystallites of Barium Titanate. - J. Phys. Soc. Japan, 1966, V.21, N 9, P.1633-1639.

[218] Demianov V.V., Soloviev S.P. Dynamical polarization and losses in ferroelectrics. - Zh. Eksp. Teor. Fiz., 1968, V.54, N 3, P.1542-1553.

[219] Turik A.V. Elastic, piezoelectric and dielectric properties of $BaTiO_3$ monocrystals with the laminated domain structure. - Fiz Tverd. Tela (Leningrad), 1970, V.12, N 3, P.892-899.

[220] Fedosov V.N., Sidorkin A.S. Dependence of the character of the domain boundaries displacement on the defect concentration. - Fiz Tverd. Tela (Leningrad), 1976, V.18, N 7, P.1967 - 1970.

[221] Turik A.V., Chernobabov A.I. On orientational contribution to the dielectric, piezoelectric and elastic constants of ferroceramics. – Zh. Tekh. Fiz., 1977, V.47, N 9, P.1944-1948.

[222] Gridnev S.A., Darinskii B.M., Fedosov V.N., Influence of domain walls dynamics on dielectric permittivity of ferroelectrics in the vicinity of Curie point. – Physics and chemistry of material processing. - 1979, N 1, P.117-120.

[223] Arlt G., Dederichs H. Complex elastic, dielectric and piezoelectric constants by domain wall damping in ferroelectric ceramics. - Ferroelectrics, 1980, V.29, N 1-2, P.47-50.

[224] Kopvillem U.Kh., Prants S.V. Domain theory of polarization echoes in ferroelectrics. - J.Physique, 1982, V.43, P.567-574.

[225] Darinskii B.M., Sidorkin A.S. Concentration of electric fields in polydomain ferroelectrics. - Fiz Tverd. Tela (Leningrad), 1984, V. 26, N.6, P.1634-1639 [Sov. Phys. Solid State, 1984, V.26, N 6, P.992-995].

[226] Bondarenko E.I., Topolov V.Yu., Turik A.V. The effect of 90° domain wall displacements on piezoelectric and dielectric constants of perovskite ceramics. - Ferroelectrics, 1990, V.110, Pt B, P.53-56.

[227] Bondarenko E.I., Topolov V.Yu., Turik A.V. The role of 90° domain wall displacements on forming physical properties of perovskite ferroelectrics. - Ferroelectrics Lett.Sec., 1991, V.13, N 1, P.13-19.

[228] Sidorkin A.S. The domain mechanism of aging and degeneration of ferroelectric material. - Proc. 9-th IEEE Intern. Symp. on Applications of Ferroelectrics. Penn. State University, IEEE Catalog Number 94 CH 3416-5, 1994, P.91-94.

[229] Darinskii B.M., Sidorkin A.S., Lazarev A.P. The elastic pliability of polydomain ferroelastics with defects. - Izv. Akad. Nauk SSSR, Ser.Fiz., 1989, V.53, N 7, P.1265-1266.

[230] Vonsovskii S.N. Magnetism [in Russian]. Moscow: Nauka, 1971. 786 p.

[231] D'yachuk P.P., Ivanov A.A., Lobov I.V. The motion of the buckling domain boundary in the continuous field of potential defects. - Izv. Akad. Nauk SSSR, Ser.Fiz., 1981, V.45, P.1701-1703.

[232] Kamysheva L.N., Zhukov O.K., Burdanina N.A., Krivitskaya T.P. Nonlinear properties of KDP crystal in strong fields. - Izv. Akad. Nauk SSSR, Ser.Fiz., 1967, V.31, N 7, P.1180-1183.

[233] Kamysheva L.N., Burdanina N.A., Zhukov O.K. On dielectric properties of KDP crystals with chromium dopants. - Kristallografiya, 1969, V.14, N 5, C.940-941.

[234] Turik A.V., Sidorenko E.N., Shevchenko N.B., Zhestkov V.F., and Proskuryakov B.F. Impact of external influences on UHF dielectric permittivity of barium titanate monocrystals. - Izv. Akad. Nauk SSSR, Ser.Fiz., 1975, V.39, N 4, P.850-853.

[235] Cross L.E., Cline T.W. Contributions to the dielectric Response from charged Domain walls in Ferroelectric $Pb_5Ge_3O_{11}$. - Ferroelectrics, 1976, V.11, P.333-336.

[236] Smits Jan G. Influence of Moving Domain Walls and Jumping Lattice Defects on Complex Material Coefficients of Piezoelectrics. – IEEE Transactions on sonics and ultrasonics, 1976, V. SU-23, N 3, P.168-173.

[237] Sidorkin A.S., Burdanina N.A., Kamysheva L.N. Increase in the real part of the permittivity in the range of the domain structure "freezing" in deuterated crystals of the potassium dihydrogen phosphate group. - Fiz. Tverd. Tela (Leningrad), 1984, V.26, N 10, P.3172-3175 [Sov. Phys. Solid State, 1984, V.26, N 10, P.1910-1912].

[238] Kamysheva L.N., Sidorkin A.S. Dielectric relaxation in crystals of $KH_2PO_4$ group. - Ferroelectrics, 1984, V.55, N 4, P.205-208..

[239] Kamysheva L.N., Sidorkin A.S., Zinovieva I.N. Dielectric Relaxation in $KH_2PO_4$ Group Crystals. - Izv. Akad. Nauk SSSR, Ser.Fiz., 1984, V.48, N 6, P.1057-1060.

[240] Kamysheva L.N., Sidorkin A.S., Milovidova S.D. Dielectric Relaxation in Ferroelectric TGS. - Physica Status Solidi, (a), 1984, V.84, K115 - K120.

[241] Shilnikov A.V., Galiyarova N.M., Nadolinskaya, Gorin S.V., Shuvaev M.A. Low-frequency dispersion of dielectric permeability in polydomain crystals of ordinary and deuterated Rochelle salt in the vicinity phase transition points. - Kristallografiya, 1986, V.31, N 2, P.326-332.

[242] Kamysheva L.N., Burdanina N.A., Zhukov O.K., Darinskii B.M., Sizova L.N. On dielectric properties of irradiated $KH_2PO_4$ crystals. - Izv. Akad. Nauk SSSR, Ser.Fiz., 1970, V.34, P.2612-2615.

[243] Burdanina N.A., Kamysheva L.N., Drozhdin S.N. Influence of the dose power of radiation on the properties of $KH_2PO_4$ crystals. - Kristallografiya, 1972, V.17, N 6, P.1171-1174.

[244] Peshikov E.V. Radiation effects in ferroelectrics [in Russian]. - Tashkent: Fan, 1986. - 139 p.

[245] Peshikov E.V. Structural sensitivity of ferroelectric phase transition and dielectric properties of the potassium dihydrogen phosphate crystals irradiated by rapid neutrons. - Kristallografiya, 1971, V.16, N 5, P.947-951.

[246] O'Keefe M., Perrino C.T. Proton conductivity in pure and doped $KH_2PO_4$. - J. Phys. Chem. Sol., 1967, V.28, N 2, P.211-218.

[247] O'Keefe M., Perrino C.T. Isotope effects in the conductivity of $KH_2PO_4$ and $KD_2PO_4$. - J. Phys. Chem. Sol., 1967, V.28, N 6, P.1086-1088.

[248] Fairall C.W., Reese W. Electrical conductivity of $KH_2AsO_4$ and $KD_2AsO_4$. - Phys. Rev. B.: 1973, V.8, N 7, P.3475-3478.

[249] Dalal N.S., Hebden J.A., McDowell C.A. EPR studies of X-ray irradi-

ated $KH_2AsO_4$ and $KH_2PO_4$-type ferroelectrics and antiferroelectrics. - J.Chem.Phys., 1975, V.62, N 11, P.4404-4410.

[250] Tokumaru Y., Abe R. Effect of irradiation on the Dielectric Constants of $KH_2PO_4$. - Japan J.Appl.Phys., 1970, V.9, N 12, P.1548-1549.

[251] Gridnev S.A., Popov V.M., Shuvalov L.A., Nechaev V.N. Temperature dependence of threshold field for dielectric losses in triglycine sulfate in the vicinity of Tc. – Fiz. Tverd. Tela (St. Peterburg), 1985, V.27, P.3-7.

[252] Gridnev S.A., Darinskii B.M., Popov V.M., Shuvalov L.A. Amplitude dependence of dielectric loss in real triglycine sulfate crystals. - Fiz. Tverd. Tela (St. Peterburg), 1986, V.28, P.2009-2014.

[253] Stankowska J., Czarnecka A. Invesitgatin of the ageing process and domain structure of TGS - group crystals. J. Material Science, 1993, V.28, P.4536-4543.

[254] Jiang Q.Y.,Cross L.E. Effect of porosity on electric fatique behaviour in PLZT and PZT ferroelectric ceramics. - J.Material Science, 1993, V.28, P.4536-4543.

[255] Jiang Q.Y., E.C.Subbarao, Cross L.E. Fatique in PLZT: acoustic emission as a discriminator between microcracking and domain switching. - Ferroelectrics, 1994, V.154, P.1251-1256.

[256] Jiang Q.Y., E.C.Subbarao, Cross L.E. Effect of electrodes and electroding methods on fatique behavior in ferroelectric materials. - Ferroelectrics, 1994, V.154, P.1257-1262.

[257] Brennan C.J., Parrella R.D., Larsen D.E. Temperature dependence fatique rates in thin-film ferroelectric capacitors. - Ferroelectrics, 1994, V.151, P.33-38.

# Index

## K



## L



## M



## N



## O



## P



## Q



## R



## S



## T



## U



## V



## Z